Ubiquitous Point Cloud

Point clouds from LiDAR and photogrammetry are vital sources of geospatial information besides remote sensing imagery. This book provides the latest theory and methodology for point cloud processing with AI to better serve earth observation, 3D vision, autonomous driving, smart city, and geospatial information applications. It covers various aspects of 3D geospatial information, including data capturing, fusing, geocomputing, modeling, and vast downstream applications. With the inclusion of numerous illustrations, diagrams, and practical applications, readers will better understand the point cloud, and its technical challenges, and learn how to utilize point cloud in different fields.

Features

- Provides in-depth point cloud processing pipeline, cutting-edge theory, and technology with AI.
- Includes many specific applications of point clouds in the geospatial field.
- Offers a comprehensive step-by-step guide from theory to application in point cloud processing.
- Includes ample supplementary materials including datasets, tools, and other online resources.
- Helps readers across many disciplines, from geospatial to engineering, to understand the vast application of point clouds and further inspires new ideas and innovative thoughts.

This book is an excellent resource for researchers, academics, students, and professionals in a variety of fields including Geomatics, Remote Sensing, Cartography and Geographic Information Systems, Data Science, Geography, Earth Science, and more.

Imaging Science Series

Qihao Weng
Hong Kong Polytechnic University

Imaging science is a multi-disciplinary field concerned with the generation, collection, duplication, analysis, modification, and visualization of images, including imaging things that the human eye cannot detect. As an evolving field, it includes research and researchers from physics, mathematics, electrical engineering, computer vision, computer science, remote sensing, and perceptual psychology. The book series will review the sub-fields within imaging science that include: image processing, computer vision, 3D computer graphics, animations, atmospheric optics, astronomical imaging, digital image restoration, digital imaging, color science, digital photography, holography, magnetic resonance imaging, medical imaging, microdensitometry, optics, photography, remote sensing, radar imaging, radiometry, silver halide, ultrasound imaging, photoacoustic imaging, thermal imaging, visual perception, and various printing technologies. The books in this series assume that readers have a basic knowledge of remote sensing and imaging science, and therefore will not cover very basic materials but concentrate on those aspects of imaging science and the application discipline(s) that are pertinent for a 'how to' knowledge. This provides academic authors and leading professionals an opportunity to combine their research, consultancy, and interesting project work into a book that is then relevant to professional lines of work or teaching modules focusing on imaging science applications.

High Spatial Resolution Remote Sensing
Data, Analysis, and Applications
Edited by Yuhong He and Qihao Weng

Remote Sensing Time Series Image Processing
Qihao Weng

Handbook of Geospatial Approaches to Sustainable Cities
Edited by Qihao Weng, in collaboration with Cheolhee Yoo

Ubiquitous Point Cloud
Theory, Model, and Applications
Bisheng Yang, Zhen Dong, Fuxun Liang, and Xiaoxin Mi

www.routledge.com/Imaging-Science/book-series/CRCIS

Ubiquitous Point Cloud
Theory, Model, and Applications

Bisheng Yang
Zhen Dong
Fuxun Liang
Xiaoxin Mi

CRC Press is an imprint of the
Taylor & Francis Group, an **informa** business

Designed cover image: © Bisheng Yang, Zhen Dong, Fuxun Liang, Xiaoxin Mi

First edition published 2025
by CRC Press
2385 NW Executive Center Drive, Suite 320, Boca Raton FL 33431

and by CRC Press
4 Park Square, Milton Park, Abingdon, Oxon, OX14 4RN

CRC Press is an imprint of Taylor & Francis Group, LLC

Library of Congress Cataloging-in-Publication Data
Names: Yang, Bisheng (Geoinformatics), author. | Dong, Zhen
(Geoinformatics), author. | Liang, Fuxun, author. | Mi, Xiaoxin, author.
Title: Ubiquitous point cloud : theory, model, and applications / Bisheng
Yang, Zhen Dong, Fuxun Liang, and Xiaoxin Mi.
Description: First edition. | Boca Raton, FL : CRC Press, 2025. | Series:
Imaging Science | Includes bibliographical references and index.
Identifiers: LCCN 2024031835 (print) | LCCN 2024031836 (ebook) | ISBN
9781032780467 (hbk) | ISBN 9781032780474 (pbk) | ISBN 9781003486060
(ebk)
Subjects: LCSH: Artificial intelligence--Environmental applications. |
Geographic information systems. | Geospatial data. | Point clouds.
Classification: LCC GE45.D37 Y36 2025 (print) | LCC GE45.D37 (ebook) |
DDC 910.285/642--dc23/eng/20241019
LC record available at https://lccn.loc.gov/2024031835
LC ebook record available at https://lccn.loc.gov/2024031836

ISBN: 978-1-032-78046-7 (hbk)
ISBN: 978-1-032-78047-4 (pbk)
ISBN: 978-1-003-48606-0 (ebk)

DOI: 10.1201/9781003486060

Typeset in Times
by SPi Technologies India Pvt Ltd (Straive)

Contents

PART II Fusion and Enhancement of Point Cloud

PART III Detection and Segmentation of Point Cloud

PART V Software and Applications

Preface

Points (x,y,z) serve as a discriminative representation in 3D (three-dimensional) digitalization and understanding of the physical world. Rapid and accurate acquisition of 3D geoinformation is a pivotal scientific challenge within Geomatics and Geoinformatics. Furthermore, it plays a significant role in advancing geological research, monitoring natural resources, enabling navigation and location-based services, facilitating urban management, supporting remote sensing applications, and enhancing intelligent transportation systems. Over the years, the means of acquiring point clouds, primarily through laser scanning and oblique photography, have advanced rapidly across aerial, terrestrial, and marine domains. New data sources, such as point clouds, continually emerge, characterized by vast data volumes, high redundancy, high density, and irregular distributions. There is an urgent need to address critical deficiencies in automation levels for object extraction and representation and intelligent knowledge-based services for specific requirements.

Systematical and theoretical approaches in intelligent point cloud processing are essential to bridge the gap between source point clouds and downstream applications. Our team has focused on three core challenges in ubiquitous point clouds for many years: data fusion and enhancement, intelligent understanding of the 3D environment, and task-specific modeling and analysis. We have conducted technical breakthroughs via engineering case studies and finally established a generalized theory of ubiquitous point clouds. This book elaborates our research efforts and accumulations on ubiquitous point cloud processing, specifically in topics such as data fusion and enhancement, 3D object detection and point cloud segmentation, as well as 3D modeling and spatial analysis. It aims to provide a systematic review of ubiquitous point clouds and serve as a valuable reference for relevant personnel.

In writing this book, we comprehensively reviewed recent research progress and breakthroughs in point cloud processing, endeavoring to present novel and accessible content. The book comprises five parts and fourteen chapters, each relatively self-contained yet logically interconnected. This ensures the overall coherence and systematized structure as follows: Part I first introduces basic concepts of ubiquitous point clouds, along with acquisition devices, processing software, and data characteristics. Then, three core components of ubiquitous point clouds theory are presented: fusion and enhancement, detection and segmentation, modeling and analysis. Following the instructions of ubiquitous point clouds theory, Part II covers theoretical methods of point cloud fusion and enhancement; Part III discusses vital points and innovative approaches in 3D object detection and point cloud segmentation; Part IV focuses on multi-detail levels of 3D modeling and analysis. Based on the theory of ubiquitous point clouds, Part V introduces the specific engineering applications and forecasts future research with great potential in ubiquitous point clouds.

Yiyao Cheng, Yige Chen, Chi Chen, Haiping Wang, Yuning Peng, Yuan Wang, Pangyin Li, Xu Han, Bo Qiu, Ruiqi Ma, Ang Jin, Yuzhou Zhou, Kunning Zheng, Ronggang Huang, Pangyin Li, Pingbo Hu, Shulin Li, Chong Liu, Xin Zhao, and

Jianping Li contributed chapters to this book. We sincerely thank them for their invaluable contributions.

The completion of this work was supported by the following projects: National Natural Science Foundation Project (No. 42130105), National Natural Science Foundation of China Outstanding Young Scientist Fund project (No. 41725005), National Natural Science Foundation Project (No. 42171431), Ministry of Education of China's Chang Jiang Scholars Program and General Program of National Natural Science Foundation of China (No. 41371431) and (No. 41071268).

While we have endeavored to ensure accuracy, inadvertent omissions may occur in this book. We sincerely invite corrections and suggestions from experts, scholars, and colleagues, which can be directed to bshyang@whu.edu.cn, dongzhenwhu@whu.edu.cn, liangfuxun@whu.edu.cn, mixiaoxin@whu.edu.cn. We extend heartfelt thanks in advance for any feedback received.

About the Authors

Dr. Bisheng Yang is a full Professor in LiDAR and photogrammetry at Wuhan University, China, and director of State Key Laboratory of Information Engineering in Surveying, Mapping and Remote Sensing (LIESMARS). His research expertise includes LiDAR and UAV photogrammetry, point cloud processing, GeoAI, GIS, and remote sensing applications. He has so far published more than 100 papers in peer-review journals, conferences and workshop proceedings. He has been Co-Chair of Point Cloud Processing Workgroup in the Photogrammetry Commission of the International Society for Photogrammetry and Remote Sensing (ISPRS) since 2016. He also serves as Associate Editor of the ISPRS Journal of Photogrammetry and Remote Sensing. He is the recipient of many academic awards, including the Carl Pulfrich Award (2019) and the Smart City Technology Innovation Award Gold Medal (2023).

Prof. Dr. Zhen Dong honorably received his B.E. and Ph.D. degrees in remote sensing and photogrammetry from Wuhan University in 2011 and 2018, respectively. After his graduation, he worked as a post-doctoral researcher at LIESMARS, Wuhan University for two years. In 2022, he was promoted to full professor at LIESMARS, Wuhan University for his outstanding works in intelligent spatial understanding. His research interest lies in the field of 3D computer vision, particularly 3D reconstruction, scene understanding, and point cloud processing, as well as their applications in intelligent transportation systems, digital twin cities, urban sustainable development, and robotics.

Dr. Fuxun Liang received his Ph.D. degree in remote sensing and photogrammetry from Wuhan University. He is currently a post-doctoral fellow in the LIESMARS at Wuhan University. His research interest lies in the field of point cloud processing and its applications in building and energy.

Dr. Xiaoxin Mi received her B.E. and Ph.D. degrees in remote sensing and photogrammetry from Wuhan University in 2016 and 2023, respectively. She is currently a post-doctoral fellow at Wuhan University of Technology. Her research interests include point cloud classification and segmentation, and road infrastructure modeling.

Part I

Why Point Cloud Matters

1 Introduction of Point Cloud

Accurate and real-time geospatial information, especially three-dimensional geospatial information, is an indispensable and important support for research on Earth system science. The representation of two-dimensional spatial data, such as maps and images, has a long history. Nonetheless, it is far from satisfying people's needs to understand real three-dimensional space and conduct geoscience research. Active acquisition methods represented by laser scanning and passive acquisition methods represented by oblique photography provide direct and effective means for the digitalization of the real world, obtaining dense point clouds with three-dimensional spatial position and attributes. Point clouds have become another vital data source after maps and images, and they play a very important role in scientific and engineering research on global change, smart cities, global mapping, intelligent transportation, and other fields. How to transform point clouds into structured and functional three-dimensional geospatial information is not only a scientific necessity but also lies at the forefront of geoscience research and engineering applications (B. Yang, Liang, and Huang 2017).

The mobile laser scanning system integrates global positioning systems, inertial measurement units, and laser scanners on different platforms to jointly calculate the position and orientation information of the laser emitter, as well as the distance to the target area, thereby obtaining a three-dimensional point cloud of the target area. Currently, point clouds can be acquired from various platforms, including satellite-based, manned/unmanned airborne-based, vehicle-based, ground-based, and portable backpack-based systems. Satellite-based laser scanning systems include the ICESat-1 and ICESat-2 satellites launched by the National Aeronautics and Space Administration (NASA) of the United States in 2003 and 2018, respectively, as well as the laser altimeter carried by China's Resource 3-02 satellite launched in 2017. Manned/unmanned aerial-based and vehicle-based mobile laser scanning systems are mainly provided by companies such as Riegl, Optech, Hexagon, Leador Space, SureStar, Hi-Target, Huace, and South Group. Depth sounders use laser beams of different wavelengths, such as blue and green, to measure the depth of water bodies and obtain underwater terrain point clouds.

The photogrammetric method utilizes specialized software for photogrammetry/oblique photogrammetry (such as INPHO, nFrame, PixelGrid, etc.) to recover the position and orientation of multi-view image data, generating dense image point clouds with color information, which provide complementary information to the laser scanning point clouds. Additionally, consumer-grade depth cameras, structured light cameras, time-of-flight (TOF) cameras, and stereo cameras are widely used to acquire 3D point clouds. Examples of hand-held devices include Apple Prime Sense[1], Microsoft Kinect[2], Intel RealSense[3], ZED[4], Bumblebee[5], and others, are extensively used for indoor point cloud acquisition.

DOI: 10.1201/9781003486060-2

1.1 POINT CLOUD ACQUISITION

1.1.1 Laser Scanning Point Cloud Acquisition

After years of development, three-dimensional laser scanning hardware has made significant progress in stability, accuracy, ease of operation, and other aspects. Representative manufacturers of three-dimensional laser scanning hardware include Riegl[6] in Austria, Leica[7] in Switzerland, Trimb1e[8] in the US, Optech[9] in Canada, FARO[10] and Ve1odyne[11] in the United States, as well as SureStar [12] and Hi-Cloud[13] in China, etc. According to the types of carrying platforms, existing laser scanning systems can be classified as: terrestrial laser scanning (TLS) systems, backpack laser scanning (BLS) systems, mobile laser scanning (MLS) systems, airborne laser scanning (ALS) systems, and satellite laser scanning (SLS) systems.

1.1.1.1 Terrestrial LiDAR Point Cloud Acquisition

The terrestrial laser scanning systems (TLS) employ scanning mirrors and servo motors to collect the three-dimensional geometric information of the Earth's surface rapidly, densely, and accurately, resulting in the acquisition of three-dimensional point clouds (Figure 1.1). With the inherent advantages of mobility, flexibility, and portability, TLS have been widely used in various fields such as engineering construction, construction supervision, landslide monitoring, cultural heritage protection, industrial facility measurement, crime scene investigation, and accident reconstruction (Zang 2018). Currently, the mainstream commercial ground-based laser scanning systems include: the VZ series[14] by Rigel of Austria, the TX series[15] by Trimble in the United States, the ScanStation series[16] by Leica in Switzerland, the PHOTON series[17] by FARO in the United States, the HS series[18] by Hi-Cloud in China, and others (see Figure 1.2). The differences in performance among various terrestrial laser scanners mainly lie in the scanning distance, frequency, accuracy, and other factors. Table 1.1 lists the important parameters of the aforementioned laser scanning systems.

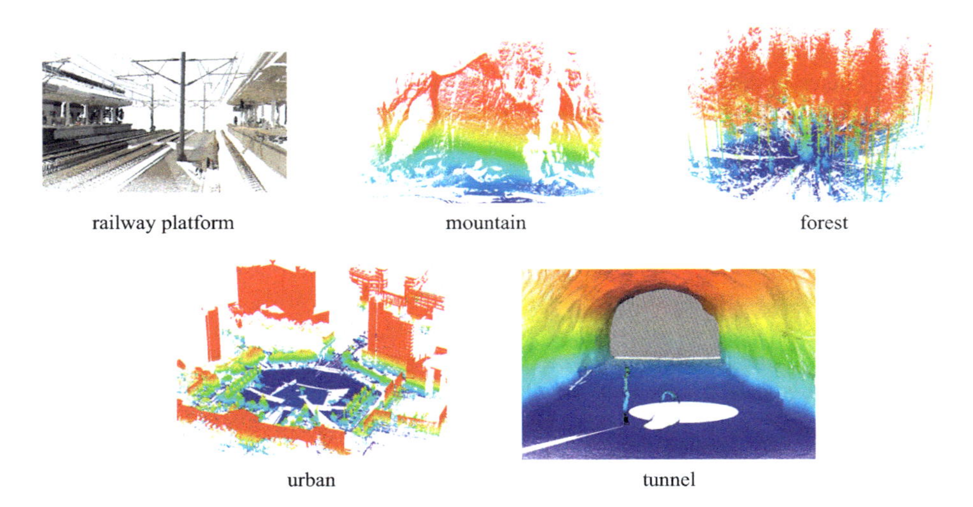

FIGURE 1.1 Typical terrestrial LiDAR point cloud scenes.

VZ-400 in Riegl

TX in Trimble

ScanStation C10 in Leica

PHOTON 120 in FARO

HS-450 in Hi-Cloud

FIGURE 1.2 Commercial terrestrial LiDAR scanner.

TABLE 1.1
The Parameters of Terrestrial LiDAR Scanner

Scanner Model		VZ 400	TX	ScanStation C10	PHOTON 120	HS 450
Manufacturer		Austria Riegl	US Trimble	Switzerland Leica	US Faro	Hi-Cloud
Scanning rate (scans/second)		122,000	500,000	50,000	976,000	500,000
Maximum range		600 m	120 m	300 m	153 m	450 m
Range accuracy		3 mm	\	4 mm	2 mm	5 mm
Field of view (FOV)	Horizontal	360°	360°	360°	360°	360°
	Vertical	100°	60°	270°	320°	100°
Ranging method		pluse	pluse	pluse	phase difference	pluse

1.1.1.2 Backpack LiDAR Point Cloud Acquisition

Hand-held/backpack laser scanning systems (BLS) integrate sensors such as laser scanners, panoramic cameras, and inertial measurement units. They utilize 3D simultaneous localization and mapping (SLAM) technology to estimate the pose of the motion platform and thereby complete the three-dimensional digitization of the environment. BLS have been widely used in forestry resources survey and management (Hyyppä et al. 2020), building information modeling (BIM) applications in construction engineering projects (Wang et al. 2018), urban 3D modeling, large-scale mapping, and other fields, which show significant advantages over vehicle-mounted mobile measurement systems and terrestrial laser scanning systems: ① they can digitize the environment in indoor and underground where global navigation satellite system (GNSS) signals are unavailable; ②they can efficiently digitize pedestrian-only areas. Currently, the mainstream hand-held/backpack laser scanning systems include the Pegasus series[19] and BLK series[20] by Leica in Switzerland, the ZEB Discovery[21], ZEB Horizon[22] and ZEB REVO[23] by GeoSLAM in the UK, the STENCIL series[24] by KAARTA in the US, the LiBackpack series[25] by GreenValley in China, the Backpacker[26] by Leador Company in China, and the 3D SLAM laser panoramic backpack robot[27] from Oslamtec in China, as shown in Figure 1.3. Figure 1.4 shows some typical point cloud scenes captured by hand-held/backpack laser scanning systems. Table 1.2 lists vital parameters of typical hand-held/backpack laser scanning systems.

FIGURE 1.3 Commercial hand-held/backpack laser scanner.

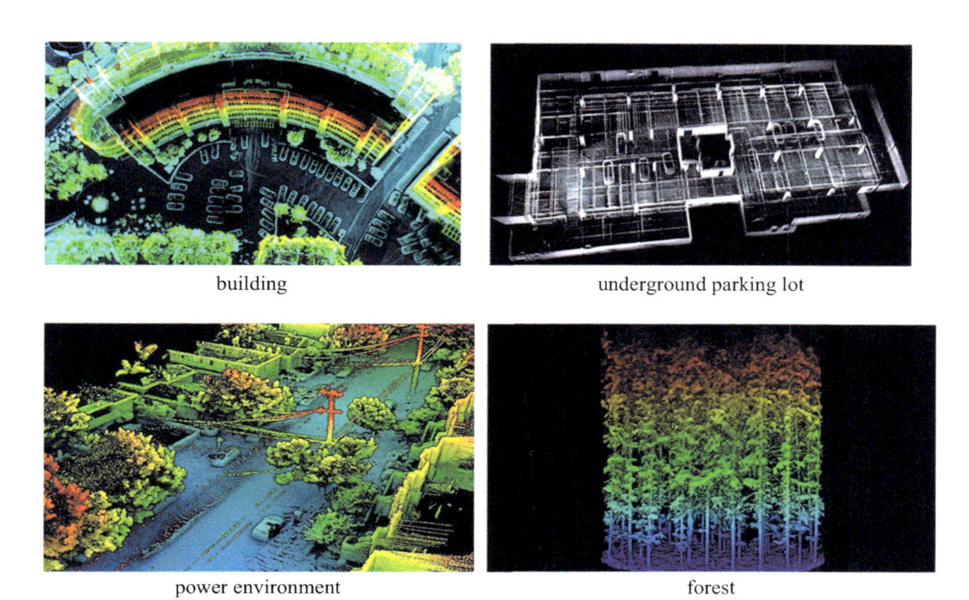

FIGURE 1.4 Point cloud obtained from a BLS system in different scenes.

1.1.1.3 Mobile LiDAR Point Cloud Acquisition

The mobile laser scanning system (MLS) installed on vehicles integrates multiple sensors such as the GNSS receiver, Inertial Navigation System (INS), laser scanner, and Charge-Coupled Device (CCD) camera. It performs trajectory calculation, calibration, and coordinate transformation of acquired images and point clouds by utilizing the positioning and attitude information provided by GNSS receiver and the IMU. This enables geolocation and generates high-precision three-dimensional coordinate information, facilitating the acquisition of three-dimensional point clouds of roads and surrounding objects.

MLS is mainly composed of the following modules: a main control module that controls the synchronization, data acquisition, recording, and transmission of various sensors; a positioning and attitude navigation module that combines GNSS receiver and IMU; a data acquisition module consisting of a laser scanner and CCD camera; and a data post-processing module that calculates point clouds and image positions, as well as performs coordinate transformation. Among them, the GNSS receiver obtains real-time geodetic coordinates of the receiver antenna center during the vehicle's motion; the IMU records the vehicle's motion attitude, including yaw, roll, and pitch angles; the laser scanner calculates the distance from the laser scanner to the measured object by measuring the time range between the emission and reflection of the laser beam; the CCD camera is mainly used to capture the texture information of objects. Due to the different installation positions of GNSS receiver, IMU, laser scanner, CCD camera, and other sensors on the vehicle, it is necessary to convert the data collected by all sensors to a unified coordinate system. This conversion allows for solving the position and attitude of each sensor relative to GNSS receiver and IMU. Furthermore, since the sensors start collecting data at different times and operate at

TABLE 1.2
The Parameters of Hand-held Laser Scanner

Scanner Model	BLK2GO	STENCIL 2-16	STENCIL 2-32	ZEB HORIZON	ZEB REVO
Manufacturer	Switzerland Leica	US KAARTA	US KAARTA	UK GeoSLAM	UK GeoSLAM
Laser scanner type	two-axis 2D laser	16-line 3D laser	32-line 3D laser	16-line 3D laser	2D laser
External rotating laser scanner	\	no	no	yes	yes
Camera type and quantity	3 global cameras (panoramic) + 1 rolling shutter camera	1 global camera	1 global camera	\	\
Maximum laser range	25 m	100 m	100 m	100 m	30 m
Point cloud relative accuracy	6–15 mm	\	\	10–30 mm	10–30 mm
Weight (excluding batteries)	650 g	1730 g	2200 g	3700 g	850 g

different data acquisition frequencies, it is also necessary to synchronize the laser scanner, IMU and CCD camera in terms of time (Wei 2012).

The mobile laser scanning system originated from the mobile highway infrastructure maintenance system in North America in 1983, which was a vehicle-mounted surveying system used for highway facility inspection. In 1988, the University of Calgary in Canada developed the Alberta MHIS system (Schwarz et al. 1993), which mounted cameras on cars and used devices such as gyroscopes, accelerometers, and odometers for positioning and orientation. It employed relative positioning methods to solve the coordinates of the positioning points and established the prototype of the on-board mobile surveying system. With the development of GPS positioning technology, in 1990, the Ohio State University developed the first mobile surveying system with modern significance, the GPSVan system (He, Novak, and Tang 1994). This system used two CCD cameras (Kodak DCS) to achieve stereo photogrammetry of roads, employed GPS, INS, and odometers for direct georeferencing of images, and calculated the three-dimensional coordinates of road points based on close-range photogrammetry principles. In 1994, the University of Calgary improved the position and orientation measurement system of Alberta MHIS and developed the first-generation VISAT system by integrating differential GPS and INS technologies (El-Sheimy 1996). Subsequently, navigation-grade INS systems were introduced, and the monochrome CCD cameras were upgraded to color CCD and video cameras, forming the second-generation VISAT system (El-Sheimy and Schwarz 1998).

With the continuous improvement of IMU performance, the advancement of GPS and INS integrated positioning technology, and the enhancements in measurement accuracy, anti-interference capability, portability and ease of operation of CCD digital sensors, laser scanners and other devices, various mobile measurement systems have been developed by various research institutions and companies. These include the TruckMap system developed by JECA in the United States (Reed, Landry, and Werther 1996), the KiSS mobile system developed by the Universität der Bundeswehr München in Germany (Klemm, Caspary, and Heister 1997), the VLSM system developed by the Center for Spatial Information Science at the University of Tokyo in Japan (Manandhar and Shibasaki 2002; H. Zhao and Shibasaki 2005), the GEOMobile system developed by the Institut Cartogràfic de Catalunya in Spain (Alamús et al. 2004), the MoSES system developed by the Helmholtz Association of German Research Centres (Graefe et al. 2001), the Roamer system in Finland (Kukko et al. 2007), the TeleAtlas system in the Netherlands, and the TiTAN system in Canada. In China, there is the WUMMS system developed by Wuhan University in 1999 (Q. Li et al. 2001), the LD2000-RM system developed by Wuhan Leador Co., Ltd. (D. Li 2006), the ZOYON-RTM system, a mobile close-range target 3D data acquisition system jointly developed by Shandong University of Science and Technology and Wuhan University (Lu et al. 2003), the 3DRMS system jointly developed by Nanjing Normal University and Wuhan University (Liang 2009), the "GPS/Beidou dual-star guidance high-dimensional real scene acquisition system" (ECNC-VLS) developed by East China Normal University (Wu et al. 2013), and the SSW-MMTS, the first mobile laser scanning system jointly developed by the Chinese Academy of Surveying and

Mapping, Capital Normal University, and other institutions in 2011. SSW-MMTS uses a domestic omnidirectional laser scanner RA-360 (D. Zhang et al. 2012). Figures 1.5 to 1.7 show the typical land-based and shipborne mobile laser scanning systems and the collected point cloud scenes. Table 1.3 lists some parameters of typical mobile laser scanning systems around the world (Kaartinen et al. 2012; Puente et al. 2011; Ellum and El-Sheimy 2002).

FIGURE 1.5 Lynx SG1 Mobile Mapper.

FIGURE 1.6 Applandix landmark Marine.

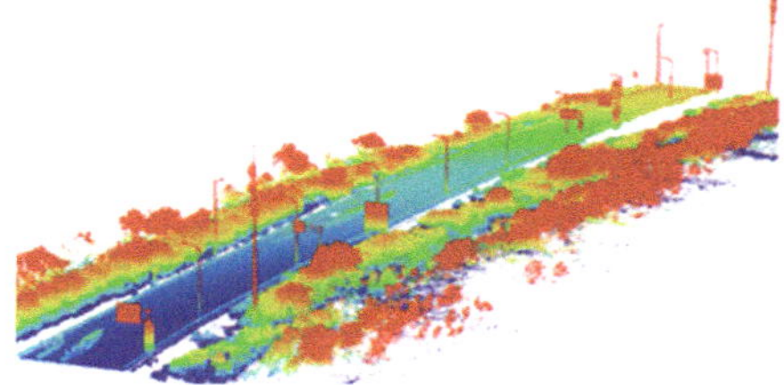

FIGURE 1.7 Point cloud obtained by MLS.

1.1.1.4 Airborne LiDAR Point Cloud Acquisition

The airborne laser scanning system, similar to the mobile laser scanning system, consists of components such as the laser scanner, GNSS receiver, IMU, and high-resolution digital camera. The airborne laser scanning system is installed on various low-, medium-, and high-altitude aircraft platforms (such as airplanes, helicopters, unmanned rotorcraft, airships, etc.) to obtain three-dimensional spatial information of the observed area. The principle for observing the coordinates of targets in the airborne LiDAR system is shown in Figure 1.8. In this process, the dynamic differential GPS determines the coordinates of the GPS antenna phase center, the IMU determines the attitude of the aircraft (pitch angle, yaw angle, and roll angle), and the laser scanner measures the distance between the laser scanning center and the observation point and records the azimuth information of the scanning mirror, thereby calculating the three-dimensional coordinates of the observation point in the WGS-84 coordinate system. Since the data measured by each observation unit (GPS, IMU, laser scanner) are in their respective independent coordinate systems, with different reference centers and coordinate axis directions. Therefore, strict coordinate system conversion is required. The coordinate system conversion involves five steps among six coordinate systems. The six coordinate systems are the instantaneous laser beam coordinate system, the laser scanning coordinate system, the inertial platform reference coordinate system, the local horizontal reference coordinate system, the local vertical reference coordinate system, and the WGS-84 coordinate system. The translation and rotation parameters for coordinate system conversion can be determined through pre-flight system calibration results (such as translation parameters of each reference center, installation angle errors, etc.) and measurement data of related sensor relationships.

Airborne laser scanning systems typically require a certain number of GPS reference stations to be set up within the observation area, along with GPS receivers on

TABLE 1.3
Typical MLS Systems Around the World

Name of MLS Systems	Research and Development Organization	POS Device	Data Acquisition Sensor	References/URLs
SSW	Capital Normal University	GPS, IMU, odometry	360-degree scanner,	(D. Zhang et al. 2012)
Lynx M1	Canada Optech	Applanix POS/LV 420	two LS, two CCD cameras	https://www.mertind.com/argentina/lidar_terrestre/Especificaciones%20Lynx%20Mobile%20Mapper.pdf
Landmark	Canada Applanix	Applanix POS MV	two CCD, video, more than one LiDAR	https://www.applanix.com/pdf/LandMark_Marine_Brochure.pdf
MX8	Trimble	Applanix POS/LV 420	Two Rigel VQ-250, six CCD cameras	http://www.hydronav.com/pdf/Trimble_MX8.pdf
VMX 450/250	Austria Rigel	GNSS, IMU	Two Riegl VQ-450/250, one VMX-450-CS6 digital camera system	http://www.riegl.com/uploads/tx_pxpriegldownloads/DataSheet_VMX-450_2015-03-19.pdf
4S-Van	South Korea ETRI and Daejeon	GPS, IMU, DMI	CCD camera	(Lee et al. 2006)
FGI Roamer	Finnish Geodetic Institute (FGI)	GPS, INS	One LiDAR, two CCD camera	(Kukko et al. 2007)
ON-SIGHT	The Ohio State University	GPS, navigation-grade IMU	five digital CCD cameras	https://www.transmap.com/services.php
VLMS	The University of Tokyo	GPS, IMU	three LiDAR, six-line scan CCD cameras	(Manandhar and Shibasaki 2002); (H. Zhao and Shibasaki 2005)
GEOMOBIL	Spain ICC	GPS, IMU	LiDAR, CCD camera	(Alamús et al. 2004)
MoSES	Unversität der Bundeswehr München	GPS, navigation-grade IMU, odometry, barometer	two CCD (compatible with LiDAR)	(Graefe et al. 2001)
WUMMS	Wuhan University	GPS, dead reckoning instrument	laser altimeter, three CCD cameras	(Q. Li et al. 2001)
TruckMAP	John E.Chance and Associates	dual frequency GPS, digital altitude sensor	non-prism laser altimeter	(Reed, Landry, and Werther 1996)

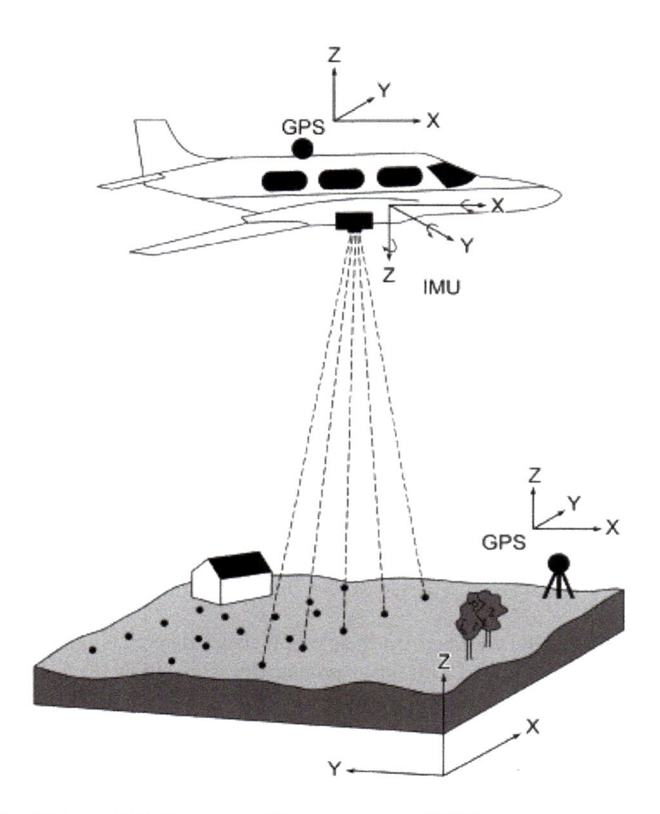

FIGURE 1.8 Airbone LiDAR system (Stateczny et al. 2023).

the aircraft, for real-time differential processing. This is done to enhance the accuracy of the aircraft's positioning. Additionally, in order to obtain more realistic texture information of the observed targets and compensate for the limitations in expressing the physical characteristics of the objects through laser data, optical imaging devices, such as CCD cameras can be installed on the aircraft. Currently, there are two main types of airborne laser scanners: multi-echo LiDAR and full-waveform LiDAR (Figure 1.9). The multi-echo LiDAR uses a simple echo detection method (e.g., constant fraction discriminator, CFD) to detect echoes in real-time and obtain the distance of the target relative to the observation center (Baltsavias 1999). The full-waveform LiDAR continuously records back-scattered signals at small time intervals (e.g., 1 ns) after the laser beam emission. By employing various waveform decomposition methods such as Gaussian decomposition (Chauve et al. 2009, 2007) and deconvolution methods, it is possible to retrieve the geometric information and physical characteristics of the observed target surface.

According to the different platforms they are installed on, airborne laser scanning systems can be divided into manned and unmanned airborne laser scanning systems (Chen 2016). Currently, the manned airborne laser scanning systems mainly include the Eclipse[28] and Galaxy[29] from Optech in Canada, the ALS series[30] and SPL series[31] from Leica in Switzerland, and the LMS-Q series[32] from Riegl in Austria, as shown in Figure 1.10.

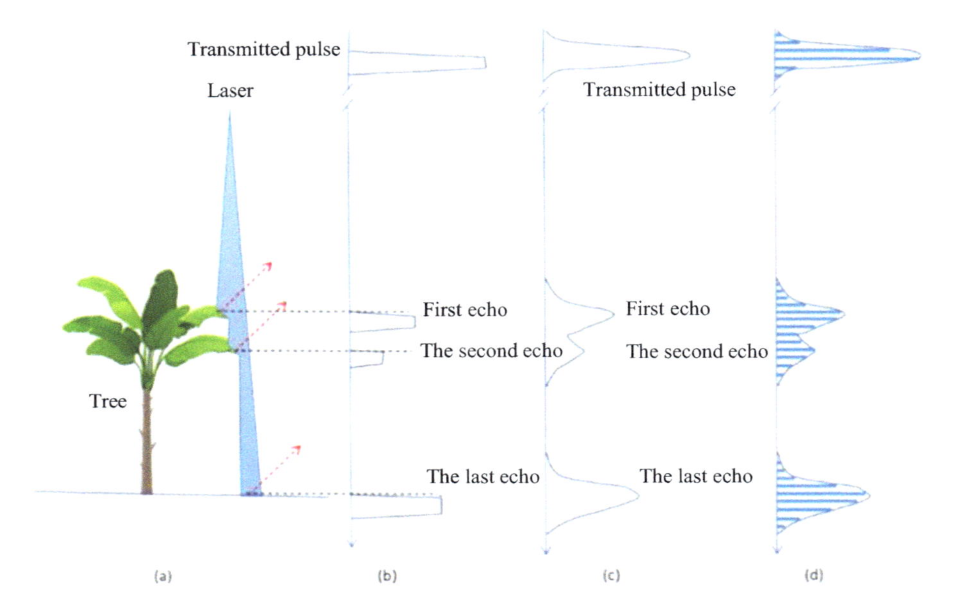

FIGURE 1.9 Diagram illustrating the principle of distance detection in multi-echo and full-waveform LiDAR systems: (a) Laser beam propagation simulation (b) Echo simulation of multi-echo system (c) Echo-simulation of full-waveform system (d) Signal discrete sampling of full-waveform system.

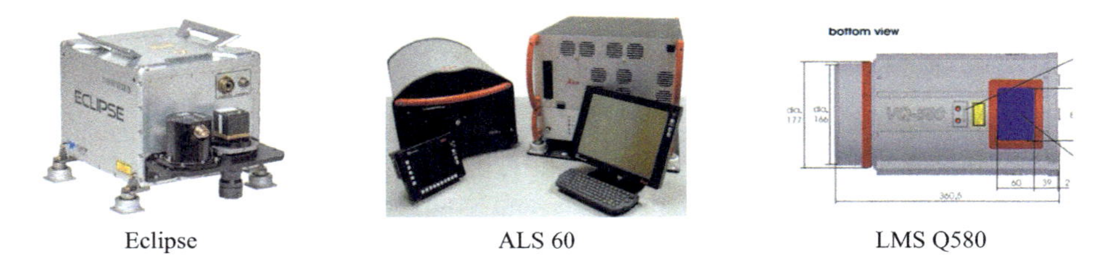

FIGURE 1.10 Manned airborne laser scanning system.

The unmanned aerial vehicle (UAV)-based mobile laser scanning system possesses advantages such as maneuverability, strong controllability, low cost, and minimal susceptibility to external environmental influences compared to traditional surveying methods. It plays an increasingly important role in industries such as basic surveying, smart cities, navigation, and location services (Chen 2016). Currently, the mainstream UAV-based mobile laser scanning systems include: the modular multi-sensor integrated mobile measurement system developed by the University of Tokyo on the Yamaha RPH2 UAV platform (Nagai et al. 2009); the Heli-mapper low-altitude unmanned aerial vehicle LiDAR system developed by the State Key Laboratory of Information Engineering in Surveying, Mapping, and Remote Sensing at Wuhan University (B. Yang and Chen 2015); the TerraLuma UAV-LiDAR system constructed by Wallace using an octocopter drone system with a lightweight LiDAR

(Wallace et al. 2012); the UAV platforms VUX-SYS[33] and miniVUX[34] introduced by Riegl, the Kylin Cloud unmanned aerial vehicle LiDAR system developed by Wuhan University (B. Yang and Li 2018), as shown in Figure 1.11. Figure 1.12 shows point clouds obtained by an airborne laser scanning system, which can be visualized in various ways according to different needs.

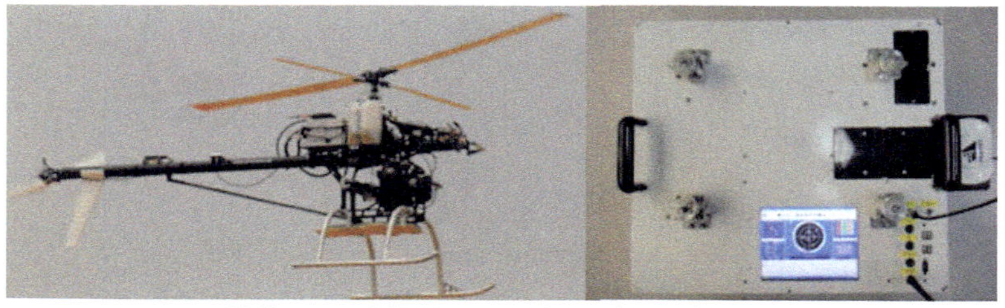

Heli-mapper LiDAR system by Wuhan University

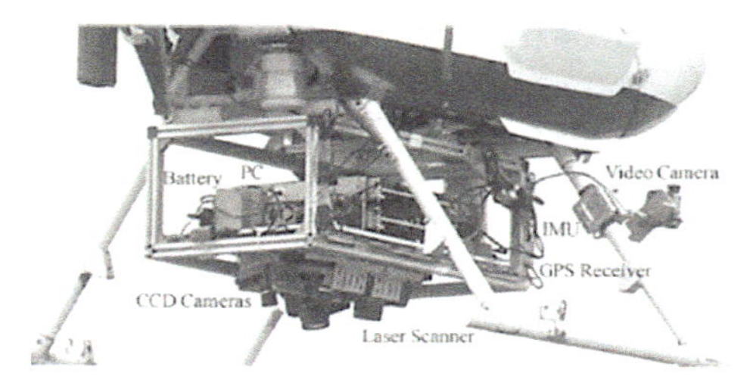

Unmanned aerial remote sensing System equipped with LiDAR multispectral camera developed by the University of Tokyo

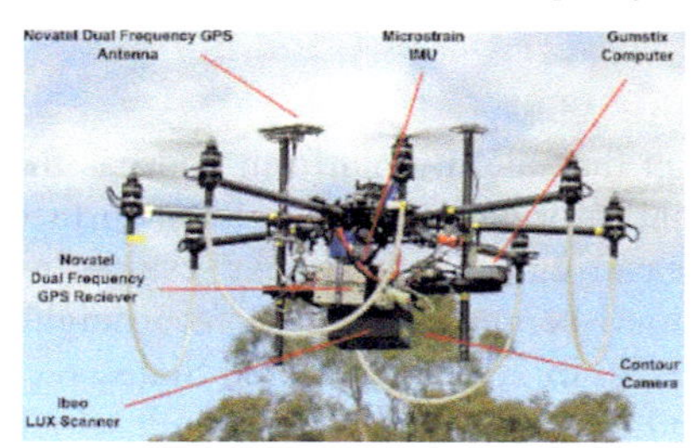

TerraLuma UAV-LiDAR system

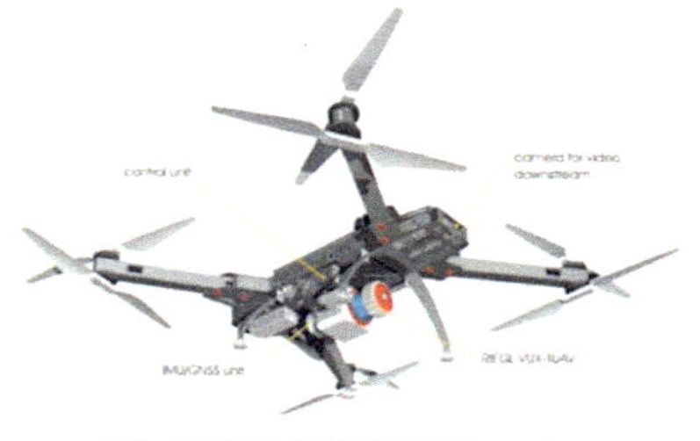

Riegl VUX-SYS LiDAR system

The Kylin-Cloud unmanned aerial vehicle LiDAR system

FIGURE 1.11 Lightweight small-scale UAV mounted mobile laser scanning system.

Global map illustration (texture)

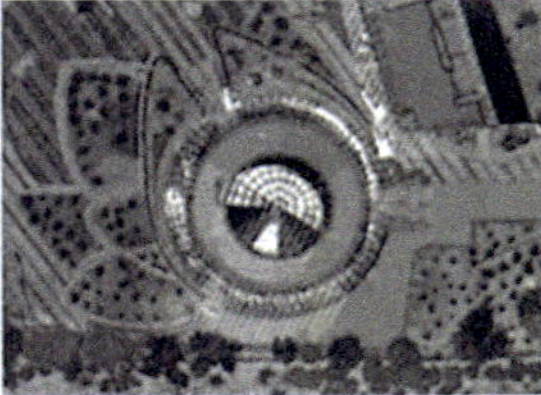

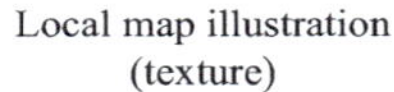

Local map illustration (texture) | Local map illustration (intensity) | Local map illustration (elevation)

FIGURE 1.12 Airborne LiDAR point cloud.

1.1.1.5 Satellite Laser Scanning Point Cloud Acquisition

The satellite laser scanning system is based on the satellite platform and has the capability to actively acquire three-dimensional information of the global surface and targets. It plays an important role in measuring polar ice sheet, estimating vegetation height and biomass, measuring cloud height and sea surface height, and monitoring global climate (G. Li et al. 2018). In 2003, the United States successfully launched the ICESat (Ice, Cloud and land Elevation Satellite)[35], which carried the Geoscience Laser Altimeter System (GLAS) for monitoring polar ice sheet, measuring sea ice elevation, estimating forest biomass, and acquiring global land elevation control point, etc, forming a broad international impact. In 2018, the United States launched ICESat-2[36], designed to serve for three years, carrying the Advanced Topographic Laser Altimeter System (ATLAS). The ATLAS laser is a visible green pulse (532 nm wavelength) divided into six beams arranged in three pairs, with each pair of laser points spaced approximately three km apart, as shown in Figure 1.13. On May 30, 2016, China successfully launched the ZY-3 02[37] satellite, carrying the first domestic earth observation laser altimeter payload. The payload is primarily used to test the performance of the laser altimeter and conduct applications such as acquiring global elevation control points, estimating tree height, and assessing forest biomass. The point cloud obtained by the ICESat-2 system is shown in Figure 1.14.

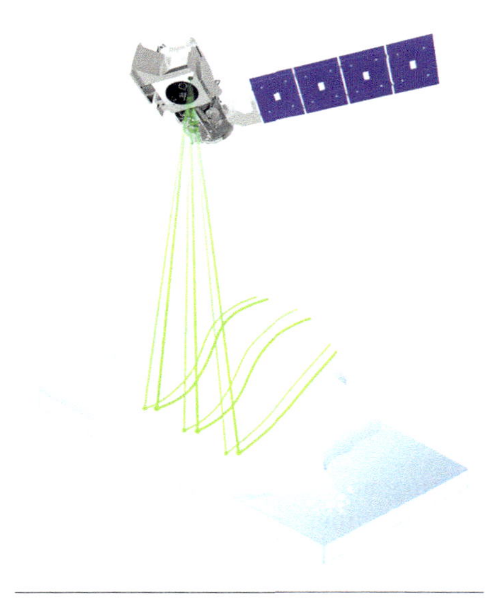

FIGURE 1.13 ICESat-2 satellite's laser scanning system.

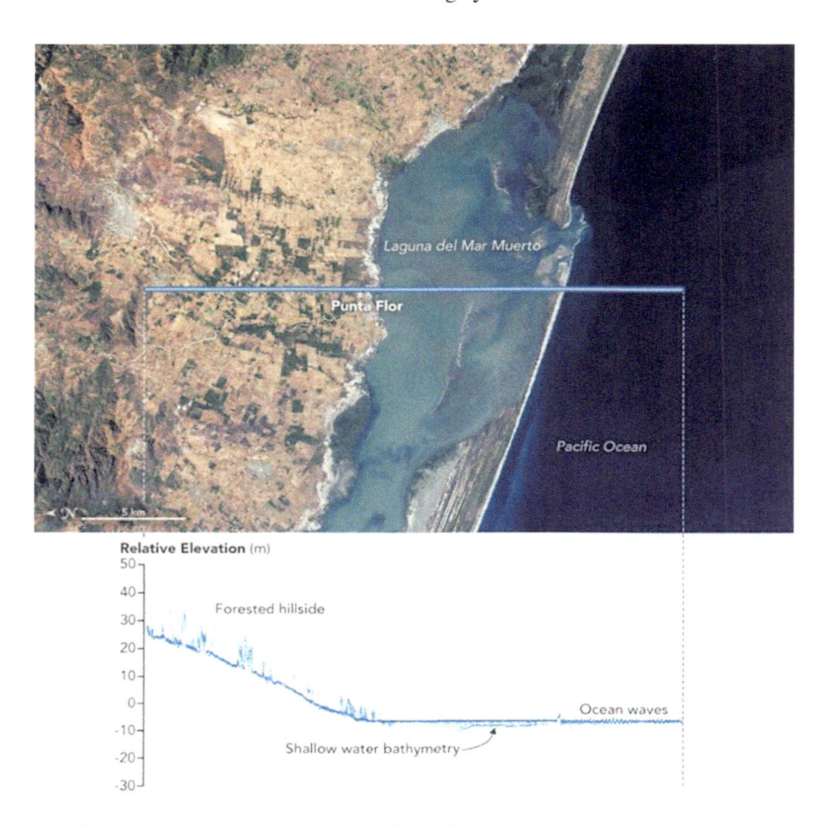

FIGURE 1.14 Point clouds obtained by ICESat-2 satellite.

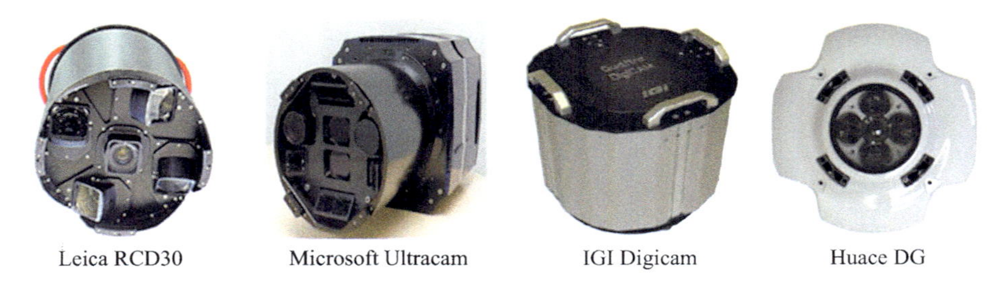

FIGURE 1.15 Photogrammetric hardware equipment.

1.1.2 Image Point Cloud Acquisition

Oblique photography/photogrammetry provides an effective means of obtaining three-dimensional point clouds from two-dimensional images. The generated point clouds have the characteristics of high density and rich texture. Point clouds derived from images have been widely used in producing digital surface models (DSMs), digital elevation models (DEMs), and digital orthophoto maps (DOMs). Currently, the main photogrammetry hardware includes Leica RCD30[38], Microsoft UltraCam[39], IGI DigiCam[40], and Huace DG3[41], as shown in Figure 1.15.

The process of generating a 3D point cloud from collected 2D images mainly includes two parts: image triangulation and dense matching. Image triangulation utilizes corresponding points matching, ground control points, airborne POS (Position and Orientation System), and other information to determine the extrinsics of the images. Dense matching performs pixel-level line matching to generate a high-density point cloud, as shown in Figure 1.16. Currently, there are many commercial photogrammetric processing softwares available, such as Photosca[42], Pix4dMapper[43], ContextCapture[44], DP-Grid, etc.

FIGURE 1.16 Image dense matching point cloud.

1.1.3 Point Cloud Acquisition by WHU-Helmet

With the rapid development of crowdsourced geographic information acquisition and simultaneous localization and mapping (SLAM), low-cost and compact wearable laser scanning (WLS) systems (Karam et al. 2019; Lowe, Kim, and Cox 2018; Ramezani et al. 2020) have become an effective complement to the traditional mobile laser scanning (MLS) systems (Hyyppä et al. 2020; Polewski et al. 2019) in complicated Global Navigation Satellite System (GNSS)-denied environments.

Compared with backpack and hand-held WLS systems, the helmet-based platform maintains the same direction as the user's view-sight and has the advantage of "what you see is what you get". The potential of this approach has been demonstrated in augmented reality (Hillers et al. 2004), virtual reality (Hoffman et al. 2006), and search&rescue (Choi et al. 2018). A helmet equipped with 3D sensors can provide real-time 3D sensing information for the operator improving the efficiency and safety of operations in difficult GNSS-denied environments (Beauregard 2006; J. Li et al. 2022). Thanks to the rapid development of sensor technology in recent years, laser scanners and IMUs are developing toward miniaturization, making the integration of compact helmet laser scanning (HLS) possible. As a result, HLS is currently an active topic and is still in the initial stage of integrating low-cost sensors in a compact helmet and utilizing the special characteristics of such sensors for high-precision real-time SLAM (J. Li et al. 2022).

Figure 1.17 is a structure diagram of the WHU-Helmet (J. Li et al. 2023). It is equipped with LiDAR, cameras, a motherboard, a power supply, and other devices. The helmet system includes four coordinate systems: the mapping frame F_{world}, the body frame F_{body}, the laser scanning frame F_{lidar}, and the camera frame F_{camera}.

The equipment can efficiently acquire point cloud data and perform mapping using SLAM algorithms (as shown in Figure 1.18). During the operation, the relative motion between the laser scanner and the measurement target leads to positional measurement errors of objects, known as motion distortion. By compensating using IMU data within the time frame of a single laser scan, the point cloud is aligned to the initial position, thereby correcting the motion. Since the measurement frequency of the laser scanner is higher than that of the IMU, the pre-integration method is employed to obtain the IMU data within the time-frame of a single LiDAR scan. Simultaneously, the platform pose provided by the IMU allows for obtaining the coordinates of objects in the world coordinate system. In LiDAR-SLAM systems, a common approach for registering frames is feature-based. Therefore, it is possible to

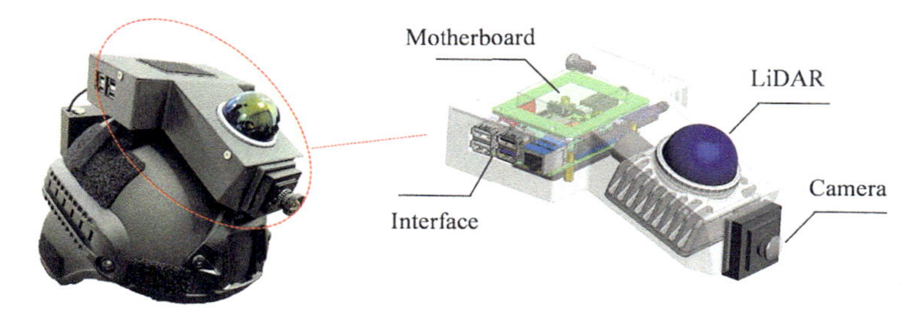

FIGURE 1.17 Mechanical structure of the WHU-Helmet.

FIGURE 1.18 The workflow of WHU-Helmet combined with SLAM algorithm for constructing a point cloud map.

use feature matching to construct an error function. Subsequently, filtering or optimization methods can be applied to iteratively solve the error function. When the results of the backend optimization meet the convergence criteria, the sub-map can be merged into the global map. The global map is continuously updated in real-time as the exploration progresses. Figure 1.19 illustrates an example of obtaining point cloud data.

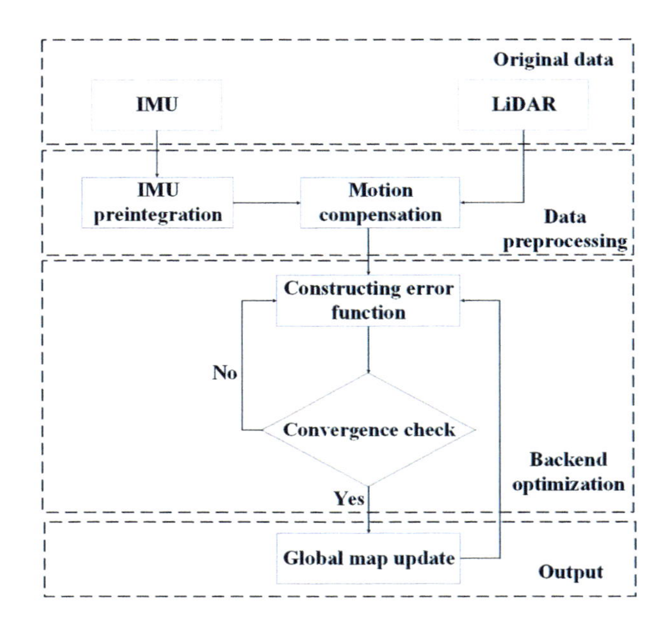

FIGURE 1.19 The example of obtaining point cloud data.

1.1.4 Trends in the Development of Point Cloud Acquisition

With the continuous increase in people's demands for the granularity and connotation of geographic spatial information, point cloud acquisition has transitioned from primarily geometric to simultaneously acquiring geometry and spectral/texture information. For example, multi-spectral laser scanning systems (Virtanen et al. 2017) enables the acquisition of both geometric and spectral information. In terms of hardwares, there has been a shift from scanning-based 3D imaging to area-array single-photon/quantum 3D imaging. Area-array single-photon LiDAR has wide application prospects in remote sensing and has become a trend in the development of active laser-based earth observation (J. Li et al. 2018). For platforms, there has been a transition from specialized equipment to diversified consumer-grade intelligent equipment. With further miniaturization, light-weighting, low-cost sensors, consumer-grade and portable integrated intelligent scanning equipment has flourished (Jiang, Zhong, and Xie 2019; D. Li 2017). The Defense Advanced Research Projects Agency (DARPA) of the US has developed autonomous cooperative scanning systems for ground robots and aerial robots. With the support of SLAM and robot control planning, these systems can scan unknown environments, greatly reduce labor costs, and solve the problem of tasks that cannot be performed manually in dangerous and specialized environments (Kelly et al. 2006).

1.1.4.1 Full-waveform LiDAR

The full-waveform LiDAR system is capable of recording the back-scattered energy of different elevation points within a certain elevation range in the form of waveforms. By analyzing the magnitude of the reflected energy at different elevation points, it can detect the vertical distribution of targets during the propagation of the laser beam (X. Yang et al. 2017; Q. Zhao, Li, and Li 2015). Compared to traditional LiDAR systems, the full-waveform LiDAR system has two advantages. Firstly, the system captures the complete echo information, which means it can provide more detailed vertical information, especially in forested areas. Secondly, by processing and decomposing received waveforms, it is possible to obtain waveform amplitudes (also known as density) and pulse widths. These two characteristics can reflect the spatial relationship and back-scattering properties of the Earth's surface, providing more useful information for object classification.

1.1.4.2 Multi-spectral LiDAR

The majority of current laser scanning systems operate in a single-wave mode, enabling the rapid acquisition of three-dimensional spatial information and the intensity of single-wave returns. While they have prominent advantages in acquiring three-dimensional spatial information, they suffer from significant limitations in discerning the physical properties of objects and detecting the states of objects on the ground. To further enhance the observation capabilities of laser scanning technology, many scholars have drawn upon the material detection capabilities of multi-/hyper-spectral remote sensing and have invented multi-spectral laser scanning systems. This technology enables laser scanning systems to retain high spatial resolution detection capabilities while also possessing spectral detection capabilities, thereby improving the system's ability to differentiate vegetation, soil, rocks, and other objects, as well as the growth status of vegetation and the composition of soil and other object states (Song S. 2015).

1.1.4.3 Single-photon LiDAR

Currently, most three-dimensional laser-based active imaging techniques primarily use scanning methods to obtain the three-dimensional information of the target by scanning the target area point by point. To achieve this, a scanning device needs to be added to the laser emission and reception system. Therefore, the system becomes large and heavy, with increased power consumption, which is detrimental to the overall miniaturization and integration of the system, while also limiting the system's resolution and sampling frequency (Luo 2015). In recent years, many countries have vigorously conducted research on non-scanning laser-based active three-dimensional imaging technology. Compact detection arrays with high integration and sensitivity, such as single-photon active three-dimensional imaging technology, can meet the requirements of system miniaturization, integration, and high-speed imaging. They also have advantages such as high sensitivity, high precision, and high resolution, making them one of the promising directions in laser-based active three-dimensional imaging technology. Currently, the mainstream commercial single-photon laser scanning systems mainly include the Sigma Space SPL[45] developed by Sigma, Space in the United States, and the SPL100 system[46] from Leica in Switzerland. The Sigma Space SPL system has high spatial resolution and sampling frequency (30 times faster than traditional scanning-based three-dimensional laser imaging technology). It can penetrate sparse vegetation, thin clouds, and water bodies, and allowing for measurements of both the surface and underwater terrains.

The diversity of point cloud acquisition platforms and methods leads to significant differences in point cloud sampling granularity, quality, representation, etc. Single platform-oriented point cloud processing methods cannot effectively coordinate multiple platform point clouds to achieve complementary advantages. There is an urgent need to develop theoretical methods for intelligent point cloud processing to provide scientific tools and means for intelligent understanding of point cloud big data.

1.2 POINT CLOUD PROCESSING

1.2.1 Automatic Point Cloud Quality Improvement

The main components of improving point cloud quality include point cloud position correction, point cloud reflectance intensity correction, and integration of point cloud attributes. In terms of position correction, methods such as strip adjustment between different point cloud strips, point position correction based on control points or fused image references, and point cloud retriangulation are used to reduce inconsistencies in the point cloud (Lichti et al. 2019; Yan et al. 2018; Cheng et al. 2018; Z. Zhang and Tao 2017; Habib et al. 2009). In terms of reflectance intensity correction, the research spotlight is on intensity calibration of airborne point clouds (Tan, Cheng, and Zhang 2017; Kashani et al. 2015; Kaasalainen et al. 2008). By calibrating the intensity, significant improvement in point cloud classification accuracy can be achieved. Regarding the integration of point cloud attribute it is mainly achieved through the fusion of point clouds and images, which enriches the attributes and enables the generation of textured point clouds (Zang et al. 2019; Chen et al. 2018; J. Li et al. 2018; B. Yang et al. 2015).

1.2.2 Accurate Description of Point Cloud Features

Point cloud feature description is a key factor in characterizing the structural morphology of point clouds, and it serves as the foundation and prerequisite for various applications such as multi-platform point cloud registration (Weber, Hänsch, and Hellwich 2015; Weinmann et al. 2015; Theiler, Wegner, and Schindler 2014), semantic information extraction (Hackel, Wegner, and Schindler 2016; Savelonas, Pratikakis, and Sfikas 2016; B. Yang et al. 2015, B. Yang and Dong 2013; Guo et al. 2013; Mian, Bennamoun, and Owens 2006), structured model reconstruction (Liu et al. 2013), and SLAM (H. Dong and Barfoot 2013; Ji Zhang and Singh 2014; Tong and Barfoot 2013). Currently, there are mainly two approaches to constructing point cloud feature descriptors: manually designed and deep network learning. In terms of manually designed features, notable example include spin images (Johnson and Hebert 1999), eigenvalue-based descriptors (Lalonde et al. 2006), fast point feature histograms (Rusu, Blodow, and Beetz 2009), rotational projection statistical feature descriptions (Guo et al. 2013), binary shape contexts (Z. Dong et al. 2017), and so on. However, these types of features rely on prior knowledge from the designer and are often sensitive to parameters. Deep learning-based methods automatically learn feature representations from a large amount of training data, and the learned features can contain thousands or even millions of parameters, thereby improving feature description capabilities (L. Zhang and Zhang 2017; Jixian Zhang, Lin, and Liang 2017). They can be divided into three categories: voxel-based, multi-view-based, and irregular point-based. Among them, representative models include VoxNet (Maturana and Scherer 2015) for voxel-based models, Mu1tiview-CNN (Su et al. 2015) for multi-view-based models, and PointNet (Charles R. Qi et al. 2017) for irregular point-based models.

1.2.3 Point Cloud Semantic Information Extraction

Semantic information extraction is the process of identifying and extracting geographic features from point clouds (Nie et al. 2019; Charles R. Qi et al. 2017; Charles Ruizhongtai Qi et al. 2017; Oesau, Lafarge, and Alliez 2014; Puente et al. 2011), providing the underlying objects and analytical basis for high-level understanding of the scene. On the one hand, point clouds contain high-density and high-precision three-dimensional information of terrain, vegetation, bridges, buildings, and transportation infrastructure. They provide a true three-dimensional perspective and miniature of the geographic targets. On the other hand, the high density, massive volume, and interstitially discrete characteristics of point clouds, the incompleteness of three-dimensional target data in scenes, and phenomena such as overlap, occlusion, and similarity among targets pose significant challenges for semantic information extraction (B. Yang et al. 2017). In terms of semantic information extraction, there are mainly feature-based methods for point-by-point classification based on feature descriptors (Guo et al. 2016, 2015) or clustering methods (Rong Huang et al. 2019; Kang and Yang 2018; B. Yang and Huang 2017; W. Zhang et al. 2016), and deep learning-based methods (Y. Song et al. 2018; Charles R. Qi et al. 2017; B. Yang et al. 2016). Compared to deep learning methods, results of feature-based semantic information extraction methods rely on the feature descriptor's capability. Deep learning network-based methods rely on the training samples and the generalization ability of the network (Wen et al. 2019; L. Zhang and Zhang 2017). Compared to the image deep learning networks, there is still huge room for improvement in both network architecture and training datasets for point cloud deep learning networks.

1.2.4 Structured Reconstruction and Scene Understanding of Point Cloud Objects

To characterize the functionality, structure, and positional relationships of objects in point cloud scenes, it is necessary to structurally represent the objects in the point cloud scene in order to support complex computational analysis. Currently, a large amount of research focuses on various aspects such as multi-levels of detail (LoD) reconstruction of building objects, reconstruction of building facades, reconstruction of trees and extraction of diameter at breast height (DBH) parameters, high-definition road maps, and indoor 3D reconstruction. Unlike mesh-based digital surface model reconstruction, the key to structured reconstruction of objects lies in accurately extracting the three-dimensional boundaries of different functional structures, thereby transforming the discrete and unordered point cloud into a geometric primitive combination model with topology. Examples include model-driven approaches (Jarząbek-Rychard and Borkowski 2016; Xiong, Elberink, and Vosselman 2014) and data-driven approaches (Xia and Wang 2018; L. Zhang et al. 2018) for building reconstruction. Model-driven methods are limited by the completeness of the model library, while data-driven structured reconstruction is influenced by data quality and may suffer from issues such as errors in structure extraction. For airborne point clouds, Bisheng Yang and Ronggang Huang proposed an iterative method based on structural constraints for generating multi-level of detail models and used a data-driven approach to construct LoD models[47]. In the field of indoor 3D reconstruction, methods are mainly based on spatial partitioning (Mura, Mattausch, and Pajarola 2016; Oesau, Lafarge, and Alliez 2014), extraction and reconstruction based on line and surface geometric elements (Cui, Li, and Dong 2019), and constructive solid geometry methods (Xiao and Furukawa 2014). To promote research on indoor 3D reconstruction, the International Society for Photogrammetry and Remote Sensing (ISPRS) has released publicly available datasets for indoor 3D reconstruction to facilitate the comparison of reconstruction results by researchers. Due to the complexity of artificial objects, large-scale 3D reconstruction of urban scenes with complex building models still requires a significant amount of manual editing. Therefore, the pursuit of automatic generation of 3D models or minimizing manual editing is an ongoing direction of research in building reconstruction.

Table 1.4 lists the download URLs, characteristics, and other information of publicly available point cloud datasets, providing support for various research purposes.

TABLE 1.4
Publicly Available Point Cloud Datasets

Dataset Name	Acquisition Equipment	Download URLs	Dataset Application	Data Description
Semantic 3D	TLS	http://semantic3d.net/	large-scale point cloud classification	three-dimensional coordinates, reflectance intensity, and color (RGB) of each point
Robotic 3D Scan Repository	TLS	http://kos.informatik.uni-osnabrueck.de/3Dscans/	point cloud registration, model reconstruction, etc.	three-dimensional coordinates, reflectance intensity, and color (RGB) of each point
Sydney Urban Objects	MLS	http://www.acfr.usyd.edu.au/papers/SydneyUrbanObjectsDataset.shtml	matching and classification	three-dimensional coordinates and reflectance intensity of each point
IQmulus & TerraMobilita	MLS	http://data.ign.fr/benchmarks/UrbanAnalysis/	segmentation, object extraction	three-dimensional coordinates, reflectance intensity, and echo numbers of each point
Paris-rue-Madame database	MLS	https://people.cmm.minesparis.psl.eu/users/serna/rueMadameDataset.html	segmentation, object extraction	three-dimensional coordinates and reflectance intensity of each point cloud
Oakland	MLS	http://www.cs.cmu.edu/~vmr/datasets/oakland_3d/cvpr09/doc/	segmentation, object extraction	three-dimensional coordinates of each point
Paris-Lille-3D-Dataset	MLS	https://npm3d.fr/paris-lille-3d	segmentation, object extraction	three-dimensional coordinates and reflectance intensity of each point
ISPRS WG II/4	ALS	https://www.isprs.org/education/benchmarks/UrbanSemLab/3d-semantic-labeling.aspx	object extraction, model reconstruction	three-dimensional coordinates of each point
Optech Titan	ALS	http://www.teledyneoptech.com/index.php/product/titan/	land cover, shallow water depth measurement	three-dimensional coordinates and three spectral information of each point (multi-spectral LiDAR)
AHN	ALS	http://dev.fwrite.org/radar/data.html	DEM generation, change detection	three-dimensional coordinates, reflectance intensity, and color (RGB) of each point
ICESat-1	SLS	https://nsidc.org/data	global ice cover, forest detection	three-dimensional coordinates of each point

(Continued)

TABLE 1.4
(Continued)

Dataset Name	Acquisition Equipment	Download URLs	Dataset Application	Data Description
ASL dataset	portable laser scanning system	https://projects.asl.ethz.ch/datasets/doku.php?id=laserregistration:laserregistration	point cloud registration	three-dimensional coordinates and reflectance intensity of each point
WHU-TLS	TLS	https://github.com/WHU-USI3DV/WHU-TLS	point cloud registration	three-dimensional coordinates and reflectance intensity of each point
WHU-Urban3D	MLSALS	https://whu3d.com/	segmentation and classification	three-dimensional coordinates, reflectance intensity, instance and classification label (part) of each point

1.3 APPLICATIONS

With the rapid development of point cloud acquisition equipment, point clouds have increasingly become a crucial data source for acquiring three-dimensional geospatial information in scientific research and engineering applications. This includes such as Earth system science, fundamental surveying and mapping, smart cities, digital preservation of cultural heritage, autonomous driving, infrastructure security monitoring, and the film and entertainment industry.

1.3.1 Earth System Science Research

The rich three-dimensional geographical information contained in point clouds accurately depicts the morphological structure of forests, glaciers, reefs, and surrounding underwater terrains. The data provides crucial support for global forest volume and biomass estimation, global glacier mass balance, marine economic development and management, as well as maritime defense and security (Ronggang Huang et al. 2019; Nie et al. 2019; Eitel et al. 2016). By obtaining point clouds of forests, it is possible to timely and accurately understand the types, quantities, qualities, distributions, growth conditions, and dynamic changes of vegetation in forest areas. It also enables the estimation of forest geometry parameters (such as forest height, horizontal distribution of the forest canopy surface, vertical structure information, etc.) and forest terrain, providing an important basis for Earth science research (such as water and carbon cycle simulation and analysis), forest conservation, formulation of forestry policies, and economic development of forestry.

1.3.2 Smart City Construction

The scale of cities and the types of municipal facilities are rapidly increasing, and the structure of cities is undergoing significant changes. Various facilities in the urban environment, such as public utilities and road traffic, are under urban management and require refined management. Theory and methods for intelligent processing of point cloud can provide comprehensive support for digital construction, BIM, underground disaster detection, and early warning, thereby establishing a fully digital underground infrastructure and dynamic aggregation of Internet of Things data. It provides scientific management and decision-making methods for the construction of a comprehensive life cycle database, including comprehensive spatial planning, construction projects, process supervision, current status, etc., as well as fine-grained management throughout the life cycle. Overall, it supports urban comprehensive management and scientific decision-making.

1.3.3 Digital Preservation of Cultural Heritage

The digitization of cultural heritage is an important scientific method for the protection, restoration, and inheritance of cultural heritage. The digital reconstruction of cultural heritage involves the complete data collection, integration, and model representation of the geometric shape, color, posture, history, and other information of cultural heritage. It forms the basis for the digital archiving of cultural heritage.

In China, cultural heritage exhibits various types, non-contact characteristics, and immobility. At the same time, there are significant geometric differences, complex shapes, and diverse global spatial resolutions in cultural heritage, posing severe challenges to three-dimensional physical world. Point cloud, as an important representation of three-dimensional physical world, provides the most direct and crucial support for the digital sampling, virtual restoration, as well as three-dimensional reconstruction and archiving of cultural heritage. It has played an important role in the digital protection of important cultural heritages in China, such as the Mogao Grottoes in Dunhuang and the virtual restoration of the Guanyin statue.

1.3.4 Autonomous Driving

The "unmanned intelligent vehicles", also known as "autonomous vehicles" or "wheel-type mobile robots", have been recognized as one of the top 20 unimaginable inventions of the next 15 years. They utilize onboard sensors to perceive the surrounding environment and control the steering and speed of the vehicle based on information obtained from sensing. This enables the vehicle to travel safely and reliably on the road. Point clouds, as a representation of the three-dimensional driving environment, serves as a scientific data source for perception of the driving environment. It can detect and track various moving targets such as pedestrians and vehicles from the collected point clouds. In addition, it can also provide scientific support for three-dimensional mapping of the road environment, including automatic recognition and extraction of road boundaries, traffic signs, and lane markings, thereby providing information for the generation of high-definition maps.

1.3.5 Infrastructure Security Monitoring

The maintenance of the built infrastructure such as railways, subways, tunnels, underground projects, power lines, and bridges are crucial for ensuring the normal functioning of socio-economic activities. Point clouds provide an indispensable data source for the three-dimensional digital modeling of infrastructure. They serve as the prerequisite and foundation for the digital management, safety analysis, and even digital twinning of various important infrastructure. By providing accurate and effective three-dimensional information, point clouds contribute to tasks such as comprehensive inspections of road surfaces (such as detecting collapses and damages), detecting deformations in bridges and tunnels, and ensuring operational safety of infrastructure.

For example, in the inspection of power line corridors (Chen, Yang, and Peng 2015), airborne LiDAR systems are utilized to acquire three-dimensional point clouds, which are then used to calculate geometric parameters of power lines and their associated equipment. These parameters include the distance between power lines and surrounding objects, the tilt angle of power towers, and the angles of power line curves. Furthermore, three-dimensional reconstruction of power lines and related equipment facilitates a visual representation of their operational status (Lin and Zhang 2016; Lai et al. 2014). This approach effectively overcomes the drawbacks of manual inspections, such as high workload and low efficiency.

1.3.6 Film and Entertainment

Point clouds play a crucial role as the primary form of precise three-dimensional digital representation of the real world. They possess unique advantages in the design and simulation of three-dimensional scenarios such as virtual interaction, simulation, augmented reality, particle simulation, and online gaming. Point clouds can support the detailed simulation of various scales, ranging from macroscopic large-scale scene construction to microscopic particle structures. They serve as a critical foundation for three-dimensional modeling, simulation analysis, etc.

1.4 CONCLUSION

This chapter focuses on the equipments, methods, and trends in point cloud acquisition, as well as the characteristics of point clouds and key aspects of point cloud processing. It also covers basic softwares for point cloud processing and various applications of point clouds. It systematically explains the significance of this new type of three-dimensional data, point clouds, in the extraction and application of three-dimensional geographic information. The aim is to provide readers with a comprehensive understanding of the point cloud acquisition, processing, and application.

NOTES

1 https://www.apple.com/.
2 www.microsoftstore.com.
3 http://www.intel.com.
4 https://www.stereolabs.com.
5 www.flir.com/iis/machine-vision/stereo-vision.
6 http://www.riegl.com/.
7 https://leica-geosystems.com.
8 https://geospatial.trimble.com.
9 http://www.teledyneoptech.com.
10 https://knowledge.faro.com.
11 http://www.velodynelidar.com/.
12 http://www.isurestar.com/.
13 http://www.hi-cloud.com.cn/.
14 http://www.riegl.com/nc/products/terrestrial-scanning/.
15 https://geospatial.trimble.com/en/products/hardware/laser-scanning.
16 https://leica-geosystems.com/products/laser-scanners/scanners.
17 https://knowledge.faro.com/Hardware/Laser_Scanner/Photon.
18 http://www.hi-cloud.com.cn/product/39/.
19 https://leica-geosystems.com/en-us/products/mobile-mapping-systems/capture-platforms/leica-pegasus_two.
20 https://leica-geosystems.com/en-us/products/laser-scanners/scanners/blk360.
21 https://geoslam.com/solutions/zeb-discovery/.
22 https://geoslam.com/solutions/zeb-horizon/.
23 https://geoslam.com/solutions/zeb-revo/.
24 https://www.exiusa.com/item/exi_lidar/stencil-slam-lidar.
25 https://www.lidar360.com/archives/3093.html.
26 https://www.leador.com.cn/MobileMeasurements/info.aspx?itemid=118.
27 http://www.oslamtec.com/?page_id=881.

28 https://www.teledyneoptech.com/en/news/newsroom/teledyne-optech-to-present-latest-lidar-applications-and-products-at-ilmf-2018/.
29 https://www.teledyneoptech.com/en/products/airborne-survey/galaxy/.
30 https://leica-geosystems.com/-/media/files/leicageosystems/products/datasheets/leica_als80_ds.ashx?la=en.
31 https://leica-geosystems.com/products/airborne-systems/lidar-sensors/leica-spl100.
32 http://www.riegl.com/uploads/tx_pxpriegldownloads/DataSheet_LMS-Q1560_2015-03-19.pdf.
33 http://www.riegl.com/products/unmanned-scanning/ricopter-with-vux-sys/.
34 http://www.riegl.com/products/unmanned-scanning/.
35 https://icesat.gsfc.nasa.gov/.
36 https://icesat-2.gsfc.nasa.gov/.
37 https://www.cresda.com/zgzywxyyzx/wxzy/zy/article/20220224111317072551484.html.
38 https://leica-geosystems.com/-/media/files/leicageosystems/products/datasheets/leica_rcd30_oblique_ds.ashx?la=en.
39 https://www.vexcel-imaging.com/.
40 https://www.igi-systems.com/largeformat-digicam.html.
41 http://www.huace.cn/.
42 https://www.agisoft.com/.
43 https://www.pix4d.com/product/pix4dmapper-photogrammetry-software.
44 https://www.bentley.com/en/products/brands/contextcapture.
45 https://spinoff.nasa.gov/Spinoff2016/ps_6.html.
46 https://leica-geosystems.com/products/airborne-systems/lidar-sensors/leica-spl100.
47 Yang and Huang, "Automated Reconstruction of Building LoDs from Airborne LiDAR Point Clouds Using An Improved Morphological Scale Space." 14.

REFERENCES

Alamús, R., A. Baron, E. Bosch, J. Casacuberta, J. Miranda, M. Pla, S. Sànchez, A. Serra, and J. Talaya. 2004. "On the Accuracy and Performance of the GEOMÒBIL System." In *Inteernational Archives of Photogrammetry and Remote Sensing, Istanbul, Turkey*.

Baltsavias, Emmanuel P. 1999. "A Comparison between Photogrammetry and Laser Scanning." *ISPRS Journal of Photogrammetry and Remote Sensing* 54 (2–3). 83–94.

Beauregard, Stéphane. 2006. "A Helmet-Mounted Pedestrian Dead Reckoning System." In *3rd International Forum on Applied Wearable Computing 2006*, 1–11. VDE.

Chauve, Adrien, Clément Mallet, Frédéric Bretar, Sylvie Durrieu, Marc Pierrot Deseilligny, and William Puech. 2007. "Processing Full-Waveform Lidar Data: Modelling Raw Signals." *International Archives of Photogrammetry, Remote Sensing and Spatial Information Sciences* 36 (part 3): W52.

Chauve, Adrien, C. Vega, S. Durrieu, F. Bretar, T. Allouis, M. Pierrot Deseilligny, and W. Puech. 2009. "Advanced Full-Waveform Lidar Data Echo Detection: Assessing Quality of Derived Terrain and Tree Height Models in an Alpine Coniferous Forest." *International Journal of Remote Sensing* 30 (19): 5211–28.

Chen, Chi. 2016. "Robust Fusion of Multi-view High-Resolution Range Imaging and Visible Light Imaging Data." Doctoral thesis, Wuhan: Wuhan University.

Chen, Chi, Bisheng Yang, Maolin Chen, Jianping Li, Xianghong Zou, Weitong Wu, and Yiheng Song. 2018. "Automatic Registration of Vehicle-Borne Mobile Mapping Laser Point Cloud and Sequent Panoramas." *Acta Geodaetica et Cartographica Sinica* 47 (2): 215.

Chen, Chi, Bisheng Yang, and Xiangyang Peng. 2015. "Automatic Registration of Low Altitude UAV Sequent Images and Laser Point Clouds." *Acta Geodaetica et Cartographica Sinica* 44 (5): 518.

Cheng, Liang, Song Chen, Xiaoqiang Liu, Hao Xu, Yang Wu, Manchun Li, and Yanming Chen. 2018. "Registration of Laser Scanning Point Clouds: A Review." *Sensors* 18 (5): 1641.

Choi, Yukyung, Namil Kim, Soonmin Hwang, Kibaek Park, Jae Shin Yoon, Kyounghwan An, and In So Kweon. 2018. "KAIST Multi-Spectral Day/Night Data Set for Autonomous and Assisted Driving." *IEEE Transactions on Intelligent Transportation Systems* 19 (3): 934–48.

Cui, Yang, Qingquan Li, and Zhen Dong. 2019. "Structural 3D Reconstruction of Indoor Space for 5G Signal Simulation with Mobile Laser Scanning Point Clouds." *Remote Sensing* 11 (19): 2262.

Dong, Hang, and Timothy D. Barfoot. 2013. "Lighting-Invariant Visual Odometry Using Lidar Intensity Imagery and Pose Interpolation." In *Field and Service Robotics: Results of the 8th International Conference*, 327–42. Springer.

Dong, Zhen, Bisheng Yang, Yuan Liu, Fuxun Liang, Bijun Li, and Yufu Zang. 2017. "A Novel Binary Shape Context for 3D Local Surface Description." *ISPRS Journal of Photogrammetry and Remote Sensing* 130: 431–52.

Eitel, Jan UH, Bernhard Höfle, Lee A. Vierling, Antonio Abellán, Gregory P. Asner, Jeffrey S. Deems, Craig L. Glennie, Philip C. Joerg, Adam L. LeWinter, and Troy S. Magney. 2016. "Beyond 3-D: The New Spectrum of Lidar Applications for Earth and Ecological Sciences." *Remote Sensing of Environment* 186: 372–92.

Ellum, Cameron, and Nasser El-Sheimy. 2002. "Land-Based Mobile Mapping Systems." *Photogrammetric Engineering and Remote Sensing* 68 (1).

El-Sheimy, Naser M. 1996. *The Development of VISAT: A Mobile Survey System for GIS Applications*. University of Calgary Alberta.

El-Sheimy, Naser, and Klaus Peter Schwarz. 1998. "Navigating Urban Areas by VISAT—A Mobile Mapping System Integrating GPS/INS/Digital Cameras for GIS Applications." *Navigation* 45 (4): 275–85.

Graefe, Gunnar, Wilhelm Caspary, Hans Heister, Jochen Klemm, and Manfred Sever. 2001. "The Road Data Acquisition System MoSES–Determination and Accuracy of Trajectory Data Gained with the Applanix POS/LV." In *Proceedings, The Third International Mobile Mapping Symposium*, Cairo, Egypt, January, 3–5. Citeseer.

Guo, Yulan, Mohammed Bennamoun, Ferdous Sohel, Min Lu, Jianwei Wan, and Ngai Ming Kwok. 2016. "A Comprehensive Performance Evaluation of 3D Local Feature Descriptors." *International Journal of Computer Vision* 116: 66–89.

Guo, Yulan, Ferdous Sohel, Mohammed Bennamoun, Min Lu, and Jianwei Wan. 2013. "Rotational Projection Statistics for 3D Local Surface Description and Object Recognition." *International Journal of Computer Vision* 105: 63–86.

Guo, Yulan, Ferdous Sohel, Mohammed Bennamoun, Jianwei Wan, and Min Lu. 2015. "A Novel Local Surface Feature for 3D Object Recognition under Clutter and Occlusion." *Information Sciences* 293: 196–213.

Habib, Ayman, Ana Paula Kersting, Ki In Bang, and Dong-Cheon Lee. 2009. "Alternative Methodologies for the Internal Quality Control of Parallel LiDAR Strips." *IEEE Transactions on Geoscience and Remote Sensing* 48 (1): 221–36.

Hackel, Timo, Jan D. Wegner, and Konrad Schindler. 2016. "Fast Semantic Segmentation of 3D Point Clouds with Strongly Varying Density." *ISPRS Annals of the Photogrammetry, Remote Sensing and Spatial Information Sciences* 3: 177–84.

He, G., K. Novak, and W. Tang. 1994. "The Accuracy of Features Positioned with the GPSVan." *International Archives of Photogrammetry and Remote Sensing* 30: 480–86.

Hillers, Bernd, D. Aiteanu, P. Tschirner, M. Park, Axel Graeser, B. Balazs, L. Schmidt, and Otto Hahn Allee N.W. 2004. "TEREBES: Welding Helmet with AR Capabilities." *Institute of Automation, University of Bremen, and Institute of Industrial Engineering and Ergonomics*, RWTH Aachen Universty. Citeseer.

Hoffman, Hunter G., Eric J. Seibel, Todd L. Richards, Thomas A. Furness, David R. Patterson, and Sam R. Sharar. 2006. "Virtual Reality Helmet Display Quality Influences the Magnitude of Virtual Reality Analgesia." *The Journal of Pain* 7 (11): 843–50.

Huang, Rong, Danfeng Hong, Yusheng Xu, Wei Yao, and Uwe Stilla. 2019. "Multi-Scale Local Context Embedding for LiDAR Point Cloud Classification." *IEEE Geoscience and Remote Sensing Letters* 17 (4): 721–25.

Huang, Ronggang, Liming Jiang, Hansheng Wang, and Bisheng Yang. 2019. "A Bidirectional Analysis Method for Extracting Glacier Crevasses from Airborne LiDAR Point Clouds." *Remote Sensing* 11 (20): 2373.

Hyyppä, Eric, Antero Kukko, Risto Kaijaluoto, Joanne C. White, Michael A. Wulder, Jiri Pyörälä, Xinlian Liang, et al. 2020. "Accurate Derivation of Stem Curve and Volume Using Backpack Mobile Laser Scanning." *ISPRS Journal of Photogrammetry and Remote Sensing* 161 (March): 246–62. doi:10.1016/j.isprsjprs.2020.01.018

Jarząbek-Rychard, M., and A. Borkowski. 2016. "3D Building Reconstruction from ALS Data Using Unambiguous Decomposition into Elementary Structures." *ISPRS Journal of Photogrammetry and Remote Sensing* 118: 1–12.

Jiang, Hemin, Ruofei Zhong, and Donghai Xie. 2019. "Design and Implementation of Mobile Measurement Method for Smartphone." *Bulletin of Surveying and Mapping*, (6): 71.

Johnson, Andrew E., and Martial Hebert. 1999. "Using Spin Images for Efficient Object Recognition in Cluttered 3D Scenes." *IEEE Transactions on Pattern Analysis and Machine Intelligence* 21 (5): 433–49.

Kaartinen, Harri, Juha Hyyppä, Antero Kukko, Anttoni Jaakkola, and Hannu Hyyppä. 2012. "Benchmarking the Performance of Mobile Laser Scanning Systems Using a Permanent Test Field." *Sensors* 12 (9): 12814–35.

Kaasalainen, Sanna, Hannu Hyyppa, Antero Kukko, Paula Litkey, Eero Ahokas, Juha Hyyppa, Hubert Lehner, Anttoni Jaakkola, Juha Suomalainen, and Altti Akujarvi. 2008. "Radiometric Calibration of LIDAR Intensity with Commercially Available Reference Targets." *IEEE Transactions on Geoscience and Remote Sensing* 47 (2). 588–98.

Kang, Zhizhong, and Juntao Yang. 2018. "A Probabilistic Graphical Model for the Classification of Mobile LiDAR Point Clouds." *ISPRS Journal of Photogrammetry and Remote Sensing* 143: 108–23.

Karam, Samer, George Vosselman, Michael Peter, Siavash Hosseinyalamdary, and Ville Lehtola. 2019. "Design, Calibration, and Evaluation of a Backpack Indoor Mobile Mapping System." *Remote Sensing* 11 (8): 905.

Kashani, Alireza G., Michael J. Olsen, Christopher E. Parrish, and Nicholas Wilson. 2015. "A Review of LiDAR Radiometric Processing: From Ad Hoc Intensity Correction to Rigorous Radiometric Calibration." *Sensors* 15 (11): 28099–128.

Kelly, Alonzo, Anthony Stentz, Omead Amidi, Mike Bode, David Bradley, Antonio Diaz-Calderon, Mike Happold, Herman Herman, Robert Mandelbaum, and Tom Pilarski. 2006. "Toward Reliable off Road Autonomous Vehicles Operating in Challenging Environments." *The International Journal of Robotics Research* 25 (5–6): 449–83.

Klemm, J., W. Caspary, and H. Heister. 1997. "Photogrammetric Data Organisation with the Mobile Surveying System KiSS." In *Proceedings of the 4th Conference on Optical*, 300–308.

Kukko, Antero, Constantin-Octavian Andrei, Veli-Matti Salminen, Harri Kaartinen, Yuwei Chen, Petri Rönnholm, Hannu Hyyppä, Juha Hyyppä, Ruizhi Chen, and Henrik Haggrén. 2007. "Road Environment Mapping System of the Finnish Geodetic Institute-FGI ROAMER." *Int. Arch. Photogramm. Remote Sens. Spat. Inf. Sci* 36: 241–47.

Lai, Xudong, Dachang Dai, Min Zheng, and Yong Du. 2014. "Powerline Three-Dimensional Reconstruction for Lidar Point Cloud Data." *Journal of Remote Sensing* 18 (6): 1223–29.

Lalonde, Jean-François, Nicolas Vandapel, Daniel F. Huber, and Martial Hebert. 2006. "Natural Terrain Classification Using Three-dimensional Ladar Data for Ground Robot Mobility." *Journal of Field Robotics* 23 (10): 839–61.

Lee, Seung-Yong, Kyoung-Ho Choi, In-Hak Joo, Seong-Ik Cho, and Jong-Hyun Park. 2006. "Design and Implementation of 4S-Van: A Mobile Mapping System." *ETRI Journal* 28 (3): 265–74. doi:10.4218/etrij.06.0105.0143

Li, Deren. 2006. "Some Thoughts on Spatial Data Uncer- Tainty in GIS." *Journal of Zhengzhou Institute of Surveying and Mapping* 23 (6): 391–92.
Li, Deren. 2017. "From Geomatics to Geospatial Intelligent Service Science." *Acta Geodaetica et Cartographica Sinica* 46 (10): 1207.
Li, Guoyuan, Jiapeng Huang, Xinming Tang, Genghua Huang, Shihong Zhou, and Yanming Zhao. 2018. "Influence of Range Gate Width on Detection Probability and Ranging Accuracy of Single Photon Laser Altimetry Satellite." *Acta Geodaetica et Cartographica Sinica* 47 (11): 1487.
Li, Jianping, Weitong Wu, Bisheng Yang, Xianghong Zou, Yandi Yang, Xin Zhao, and Zhen Dong. 2023. "WHU-Helmet: A Helmet-Based Multi-Sensor SLAM Dataset for the Evaluation of Real-Time 3D Mapping in Large-Scale GNSS-Denied Environments." *IEEE Transactions on Geoscience and Remote Sensing*. IEEE.
Li, Jianping, Bisheng Yang, Chi Chen, Ronggang Huang, Zhen Dong, and Wen Xiao. 2018. "Automatic Registration of Panoramic Image Sequence and Mobile Laser Scanning Data Using Semantic Features." *ISPRS Journal of Photogrammetry and Remote Sensing* 136: 41–57.
Li, Jianping, Bisheng Yang, Yiping Chen, Weitong Wu, Yandi Yang, Xin Zhao, and Ruibo Chen. 2022. "Evaluation of a Compact Helmet-Based Laser Scanning System for Aboveground and Underground 3D Mapping." *The International Archives of the Photogrammetry, Remote Sensing and Spatial Information Sciences* 43: 215–20.
Li, Q., B. Li, J. Chen, Q. Hu, and Y. Li. 2001. "3D Mobile Mapping System for Road Modeling." In *Proceedings of The 3rd International Symposium on Mobile Mapping Technology (MMS 2001)*. Cario, Egypt.
Liang Cheng. 2009. "Research on Vehicle-based Spatial Data Acquisition System for GIS." Master's thesis, Nanjing: Nanjing Normal University.
Lichti, Derek D., Craig L. Glennie, Kaleel Al-Durgham, Adam Jahraus, and Jeremy Steward. 2019. "Explanation for the Seam Line Discontinuity in Terrestrial Laser Scanner Point Clouds." *ISPRS Journal of Photogrammetry and Remote Sensing* 154. 59–69.
Lin, Xiangguo, and Jixian Zhang. 2016. "3D Power Line Reconstruction from Airborne LiDAR Point Cloud of Overhead Electric Power Transmission Corridors." *Acta Geodaetica et Cartographica Sinica* 45 (3): 347.
Liu, Chun, Beiqi Shi, Xuan Yang, Nan Li, and Hangbin Wu. 2013. "Automatic Buildings Extraction from LiDAR Data in Urban Area by Neural Oscillator Network of Visual Cortex." *IEEE Journal of Selected Topics in Applied Earth Observations and Remote Sensing* 6 (4): 2008–19.
Lowe, Thomas, Soohwan Kim, and Mark Cox. 2018. "Complementary Perception for Handheld Slam." *IEEE Robotics and Automation Letters* 3 (2). 1104–11.
Lu, Xiushan, Qingquan Li, Wenhao Feng, Chengming Li, Ying Chen, Yifu Li, Xiaodong Han, and Fengxiang Jin. 2003. "Vehicle-Borne Urban Information Acquisition and 3D Modeling System." *Engineering Journal of Wuhan University* 36 (3): 76–80.
Luo Hanjun. 2015. "Research on Key Technologies of Single-Photon Imaging Detection." Doctoral thesis, Wuhan: Huazhong University of Science and Technology.
Manandhar, Dinesh, and Ryosuke Shibasaki. 2002. "Auto-Extraction of Urban Features from Vehicle-Borne Laser Data." *International Archives of Photogrammetry Remote Sensing and Spatial Information Sciences* 34 (4): 650–55.
Maturana, Daniel, and Sebastian Scherer. 2015. "Voxnet: A 3d Convolutional Neural Network for Real-Time Object Recognition." In *2015 IEEE/RSJ International Conference on Intelligent Robots and Systems (IROS)*, 922–28. IEEE.
Mian, Ajmal S., Mohammed Bennamoun, and Robyn A. Owens. 2006. "A Novel Representation and Feature Matching Algorithm for Automatic Pairwise Registration of Range Images." *International Journal of Computer Vision* 66: 19–40.
Mura, Claudio, Oliver Mattausch, and Renato Pajarola. 2016. "Piecewise-planar Reconstruction of Multi-room Interiors with Arbitrary Wall Arrangements." In *Computer Graphics Forum*, 35:179–88. Wiley Online Library.

Nagai, Masahiko, Tianen Chen, Ryosuke Shibasaki, Hideo Kumagai, and Afzal Ahmed. 2009. "UAV-Borne 3-D Mapping System by Multisensor Integration." *IEEE Transactions on Geoscience and Remote Sensing* 47 (3): 701–8.

Nie, Sheng, Cheng Wang, Xiaohuan Xi, Shezhou Luo, Xiaoxiao Zhu, Guoyuan Li, Hua Liu, Jinyan Tian, and Su Zhang. 2019. "Assessing the Impacts of Various Factors on Treetop Detection Using LiDAR-Derived Canopy Height Models." *IEEE Transactions on Geoscience and Remote Sensing* 57 (12): 10099–115.

Oesau, Sven, Florent Lafarge, and Pierre Alliez. 2014. "Indoor Scene Reconstruction Using Feature Sensitive Primitive Extraction and Graph-Cut." *ISPRS Journal of Photogrammetry and Remote Sensing* 90: 68–82.

Polewski, Przemyslaw, Wei Yao, Lin Cao, and Sha Gao. 2019. "Marker-Free Coregistration of UAV and Backpack LiDAR Point Clouds in Forested Areas." *ISPRS Journal of Photogrammetry and Remote Sensing* 147. 307–18.

Puente, Xose S., Magda Pinyol, Víctor Quesada, Laura Conde, Gonzalo R. Ordóñez, Neus Villamor, Georgia Escaramis, Pedro Jares, Sílvia Beà, and Marcos González-Díaz. 2011. "Whole-Genome Sequencing Identifies Recurrent Mutations in Chronic Lymphocytic Leukaemia." *Nature* 475 (7354): 101–5.

Qi, Charles R., Hao Su, Kaichun Mo, and Leonidas J. Guibas. 2017. "Pointnet: Deep Learning on Point Sets for 3d Classification and Segmentation." In *Proceedings of the IEEE Conference on Computer Vision and Pattern Recognition*, 652–60.

Qi, Charles Ruizhongtai, Li Yi, Hao Su, and Leonidas J. Guibas. 2017. "Pointnet++: Deep Hierarchical Feature Learning on Point Sets in a Metric Space." *Advances in Neural Information Processing Systems* 30: 5099–5118.

Ramezani, Milad, Yiduo Wang, Marco Camurri, David Wisth, Matias Mattamala, and Maurice Fallon. 2020. "The Newer College Dataset: Handheld Lidar, Inertial and Vision with Ground Truth." In *2020 IEEE/RSJ International Conference on Intelligent Robots and Systems (IROS)*, 4353–60. IEEE.

Reed, Morgan D., C. E. Landry, and K. C. Werther. 1996. "The Application of Air and Ground Based Laser Mapping Systems to Transmission Line Corridor Surveys." In *Proceedings of Position, Location and Navigation Symposium-PLANS'96*, 444–51. IEEE.

Rusu, Radu Bogdan, Nico Blodow, and Michael Beetz. 2009. "Fast Point Feature Histograms (FPFH) for 3D Registration." In *2009 IEEE International Conference on Robotics and Automation*, 3212–17. IEEE.

Savelonas, Michalis A., Ioannis Pratikakis, and Konstantinos Sfikas. 2016. "Fisher Encoding of Differential Fast Point Feature Histograms for Partial 3D Object Retrieval." *Pattern Recognition* 55. 114–24.

Schwarz, K. P., H. E. Martell, N. El-Sheimy, Ron Li, M. A. Chapman, and D. Cosandier. 1993. "VIASAT-A Mobile Highway Survey System of High Accuracy." In *Proceedings of VNIS'93-Vehicle Navigation and Information Systems Conference*, 476–81. IEEE.

Song Shalei. 2015. "Fundamental Principles and Key Technologies of Ground Observation Multispectral Lidar." Doctoral thesis, Wuhan: Wuhan University.

Song, Yuhang, Chao Yang, Yeji Shen, Peng Wang, Qin Huang, and C.-C. Jay Kuo. 2018. "Spg-Net: Segmentation Prediction and Guidance Network for Image Inpainting." *arXiv Preprint arXiv:1805.03356.*

Stateczny, Andrzej, Armin Halicki, Mariusz Specht, Cezary Specht, and Oktawia Lewicka. 2023. "Review of Shoreline Extraction Methods from Aerial Laser Scanning." *Sensors* 23 (11): 5331. doi:10.3390/s23115331

Su, Hang, Subhransu Maji, Evangelos Kalogerakis, and Erik Learned-Miller. 2015. "Multi-View Convolutional Neural Networks for 3d Shape Recognition." In *Proceedings of the IEEE International Conference on Computer Vision*, 945–53.

Tan, Kai, Xiaojun Cheng, and Jixing Zhang. 2017. "Correction for Incidence Angle and Distance Effects on TLS Intensity Data." *Geomatics and Information Science of Wuhan University* 42 (2): 223–28.

Theiler, Pascal Willy, Jan Dirk Wegner, and Konrad Schindler. 2014. "Keypoint-Based 4-Points Congruent Sets–Automated Marker-Less Registration of Laser Scans." *ISPRS Journal of Photogrammetry and Remote Sensing* 96: 149–63.

Tong, Chi Hay, and Timothy D. Barfoot. 2013. "Gaussian Process Gauss-Newton for 3D Laser-Based Visual Odometry." In *2013 IEEE International Conference on Robotics and Automation*, 5204–11. IEEE.

Virtanen, Juho-Pekka, Antero Kukko, Harri Kaartinen, Anttoni Jaakkola, Tuomas Turppa, Hannu Hyyppä, and Juha Hyyppä. 2017. "Nationwide Point Cloud—The Future Topographic Core Data." *ISPRS International Journal of Geo-Information* 6 (8): 243.

Wallace, Luke, Arko Lucieer, Christopher Watson, and Darren Turner. 2012. "Development of a UAV-LiDAR System with Application to Forest Inventory." *Remote Sensing* 4 (6): 1519–43.

Wang, Cheng, Shiwei Hou, Chenglu Wen, Zheng Gong, Qing Li, Xiaotian Sun, and Jonathan Li. 2018. "Semantic Line Framework-Based Indoor Building Modeling Using Backpacked Laser Scanning Point Cloud." *ISPRS Journal of Photogrammetry and Remote Sensing*, ISPRS Journal of Photogrammetry and Remote Sensing Theme Issue "Point Cloud Processing," 143 (September): 150–66. doi:10.1016/j.isprsjprs.2018.03.025

Weber, T., R. Hänsch, and O. Hellwich. 2015. "Automatic Registration of Unordered Point Clouds Acquired by Kinect Sensors Using an Overlap Heuristic." *ISPRS Journal of Photogrammetry and Remote Sensing* 102: 96–109.

Wei, Zheng 2012. "Automated Extraction of Buildings and Facades Reconstruction from Mobile LiDAR Point Clouds." Wuhan University.

Weinmann, Martin, Steffen Urban, Stefan Hinz, Boris Jutzi, and Clément Mallet. 2015. "Distinctive 2D and 3D Features for Automated Large-Scale Scene Analysis in Urban Areas." *Computers & Graphics* 49: 47–57.

Wen, Chenglu, Xiaotian Sun, Jonathan Li, Cheng Wang, Yan Guo, and Ayman Habib. 2019. "A Deep Learning Framework for Road Marking Extraction, Classification and Completion from Mobile Laser Scanning Point Clouds." *ISPRS Journal of Photogrammetry and Remote Sensing* 147: 178–92.

Wu, Bin, Bailang Yu, Wenhui Yue, Wenqi Tan, Chunling Hu, and Jianping Wu. 2013. "Method for Identifying Individual Street Trees from the Cloud Data of the Vehicle-Borne Laser Scanning Points." *Journal of East China Normal University(Natural Sciences)*, no. 2: 38–49.

Xia, Shaobo, and Ruisheng Wang. 2018. "Extraction of Residential Building Instances in Suburban Areas from Mobile LiDAR Data." *ISPRS Journal of Photogrammetry and Remote Sensing* 144: 453–68.

Xiao, Jianxiong, and Yasutaka Furukawa. 2014. "Reconstructing the World's Museums." *International Journal of Computer Vision* 110: 243–58.

Xiong, Biao, S. Oude Elberink, and G. Vosselman. 2014. "A Graph Edit Dictionary for Correcting Errors in Roof Topology Graphs Reconstructed from Point Clouds." *ISPRS Journal of Photogrammetry and Remote Sensing* 93: 227–42.

Yan, Li, Junxiang Tan, Hua Liu, Hong Xie, and Changjun Chen. 2018. "Automatic Non-Rigid Registration of Multi-Strip Point Clouds from Mobile Laser Scanning Systems." *International Journal of Remote Sensing* 39 (6): 1713–28.

Yang, Bisheng, and Chi Chen. 2015. "Automatic Registration of UAV-Borne Sequent Images and LiDAR Data." *ISPRS Journal of Photogrammetry and Remote Sensing* 101: 262–74.

Yang, Bisheng, and Zhen Dong. 2013. "A Shape-Based Segmentation Method for Mobile Laser Scanning Point Clouds." *ISPRS Journal of Photogrammetry and Remote Sensing* 81: 19–30.

Yang, Bisheng, Zhen Dong, Fuxun Liang, and Yuan Liu. 2016. "Automatic Registration of Large-Scale Urban Scene Point Clouds Based on Semantic Feature Points." *ISPRS Journal of Photogrammetry and Remote Sensing* 113: 43–58.

Yang, Bisheng, Zhen Dong, Yuan Liu, Fuxun Liang, and Yongjun Wang. 2017. "Computing Multiple Aggregation Levels and Contextual Features for Road Facilities Recognition Using Mobile Laser Scanning Data." *ISPRS Journal of Photogrammetry and Remote Sensing* 126: 180–94.

Yang, Bisheng, Zhen Dong, Gang Zhao, and Wenxia Dai. 2015. "Hierarchical Extraction of Urban Objects from Mobile Laser Scanning Data." *ISPRS Journal of Photogrammetry and Remote Sensing* 99: 45–57.

Yang, Bisheng, and Ronggang Huang. 2017. "Automated Reconstruction of Building LoDs from Airborne LiDAR Point Clouds Using An Improved Morphological Scale Space." *RemoteSensing* 9 (1): 14.

Yang, Bisheng, and Jianping Li. 2018. "Implementation of a Low-Cost Mini-UAV Laser Scanning System." *Geomatics and Information Science of Wuhan University* 43 (12): 1972–78.

Yang, Bisheng, Fuxun Liang, and Ronggang Huang. 2017. "Progress, Challenges and Perspectives of 3D LiDAR Point Cloud Processing." *Acta Geodaetica et Cartographica Sinica* 46 (10): 1509. doi:10.11947/j.AGCS.2017.20170351

Yang, Xuebo, Cheng Wang, Xiaohuan Xi, Jianlin Tian, Sheng Nie, and Jianlin Zhu. 2017. "Wavelet Transform of Gaussian Progressive Decomposition Method for Full-Waveform LiDAR Data." *Journal of Infrared and Millimeter Waves* 36 (6): 749–55.

Zang, Yufu. 2018. "Spatial Alignment of Multi-Platform Point Clouds and On-Demand 3D Modeling." *Acta Geodaetica et Cartographica Sinica* 47 (12): 1693. doi:10.11947/j.AGCS.2018.20170730

Zang, Yufu, Bisheng Yang, Jianping Li, and Haiyan Guan. 2019. "An Accurate TLS and UAV Image Point Clouds Registration Method for Deformation Detection of Chaotic Hillside Areas." *Remote Sensing* 11 (6): 647.

Zhang, Di, Ruofei Zhong, Guang-wei Li, and Kun Zhao. 2012. "3D Data Acquisition and Applications of Vehicle-Mounted Laser Scanning Systems." *Geospatial Information* 10 (1): 20–21+24.

Zhang, Ji, and Sanjiv Singh. 2014. "LOAM: Lidar Odometry and Mapping in Real-Time." In *Robotics: Science and Systems*. 2:1–9, Berkeley, CA.

Zhang, Jixian, Xiangguo Lin, and Xinlian Liang. 2017. "Advances and Prospects of Information Extraction from Point Clouds." *Acta Geodaetica et Cartographica Sinica* 46 (10): 1460.

Zhang, Liqiang, Zhuqiang Li, Anjian Li, and Fangyu Liu. 2018. "Large-Scale Urban Point Cloud Labeling and Reconstruction." *ISPRS Journal of Photogrammetry and Remote Sensing* 138: 86–100.

Zhang, Liqiang, and Liang Zhang. 2017. "Deep Learning-Based Classification and Reconstruction of Residential Scenes from Large-Scale Point Clouds." *IEEE Transactions on Geoscience and Remote Sensing* 56 (4): 1887–97.

Zhang, Wuming, Jianbo Qi, Peng Wan, Hongtao Wang, Donghui Xie, Xiaoyan Wang, and Guangjian Yan. 2016. "An Easy-to-Use Airborne LiDAR Data Filtering Method Based on Cloth Simulation." *Remote Sensing* 8 (6): 501.

Zhang, Zuxun, and Pengjie Tao. 2017. "An Overview on 'Cloud Control' Photogrammetry in Big Data Era." *Acta Geodaetica et Cartographica Sinica* 46 (10): 1238.

Zhao, Huijing, and Ryosuke Shibasaki. 2005. "A Novel System for Tracking Pedestrians Using Multiple Single-Row Laser-Range Scanners." *IEEE Transactions on Systems, Man, and Cybernetics-Part A: Systems and Humans* 35 (2): 283–91.

Zhao, Quanhua, Hongying Li, and Yu Li. 2015. "Gaussian Mixture Model with Variable Components for Full Waveform Lidar Data Decomposition and RJMCMC Algorithm." *Acta Geodaetica et Cartographica Sinica* 44 (12): 1367.

2 Ubiquitous Point Cloud

2.1 FRAMEWORK OF UBIQUITOUS POINT CLOUD

Point $(x,y,z,\ldots)$ is the minimum unit to digitalize the physical world, which may be captured from different means, such as laser scanning, photogrammetry, simulation engines, etc. Nevertheless, points from different means are associated with various characteristics, including location accuracies, attribute richness, and spatial references. To obtain meaningful information from these points, the pre-condition is to model these points in a uniform spatial reference with uniform representation. Hence, we propose the framework of ubiquitous point cloud (UPC) and model it in a formal representation as follows.

$$\text{UPC} = \bigcup_{i=1,j=1,k=1,\ldots}^{M,N,L,\ldots} \left(X_i \cup Y_j \cup Z_k \cup ..\right) = \bigcup_{q=1}^{Q} P_q$$
$$s.t.\begin{cases} \min\limits_{s,R,T} \sum_{i=1}^{M}\sum_{j=1}^{N} \left| X(\mathrm{x_i},\mathrm{y_i},\mathrm{z_i}) - s\left(RY(\mathrm{x_j},\mathrm{y_j},\mathrm{z_j}) + T\right) \right|_2 \\ \min\limits_{s,R,T} \sum_{i=1}^{M}\sum_{j=1}^{N} \left| X(\mathrm{a_i}) - Y(\mathrm{a_j}) \right| \\ A_{c_a} \cap A_{c_b} = \varnothing \end{cases} \tag{2.1}$$

where $P_q = (x_q, y_q, z_q, a_q)$ is a point belonging to UPC, and (x_q, y_q, z_q) and a_q are the three-dimensional coordinate and the associated attribute of P_q, respectively. X_i, Y_j, and Z_k are the points captured from various means or platforms. The UPC is necessary to meet the geometric and attribute constraints, where (s, R, T) are the geometric transformation parameters. Additionally, attributes of points with different semantic labels c_a and c_b should be distinguishable.

The core components of the ubiquitous point cloud are shown in Figure 2.1.

Regarding point cloud *fusion and enhancement*, Prof. Bisheng Yang first proposed the scientific concept and theoretical research framework of "ubiquitous point cloud" (Yang and Wang 2016). It was selected by the International Society of Photogrammetry and Remote Sensing as one of the vital research topics and initially established the theory and framework of the ubiquitous point cloud. This provides scientific support for the intelligent processing and engineering application of multi-platform point clouds. It has been highly recognized by international peers and was awarded the unique Carl Pulfrich Award in 2019. The ubiquitous point cloud theory initially defines the minimum descriptive unit of the objective natural world, which is a point. The points in the ubiquitous point cloud are not just simple geometric points in space but have several essential characteristics, including spatial position, boundary, time,

DOI: 10.1201/9781003486060-3

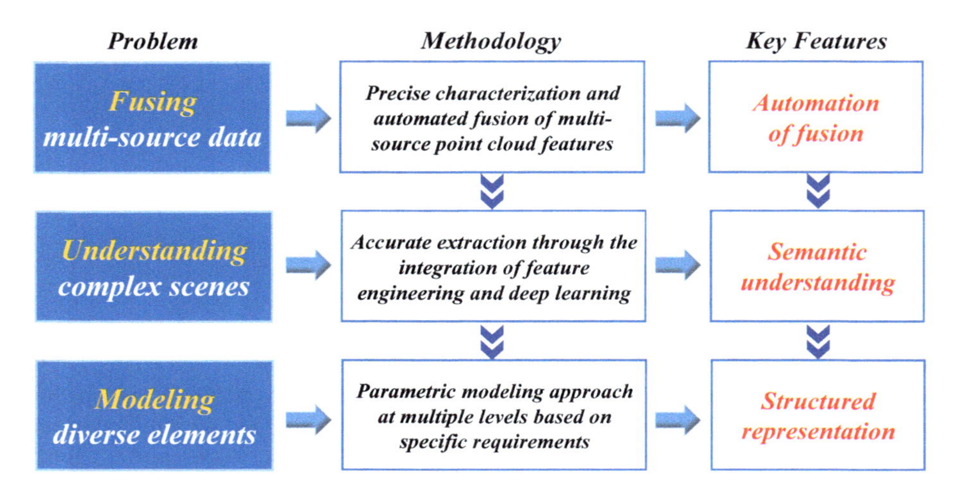

FIGURE 2.1 The core components of the ubiquitous point cloud theory.

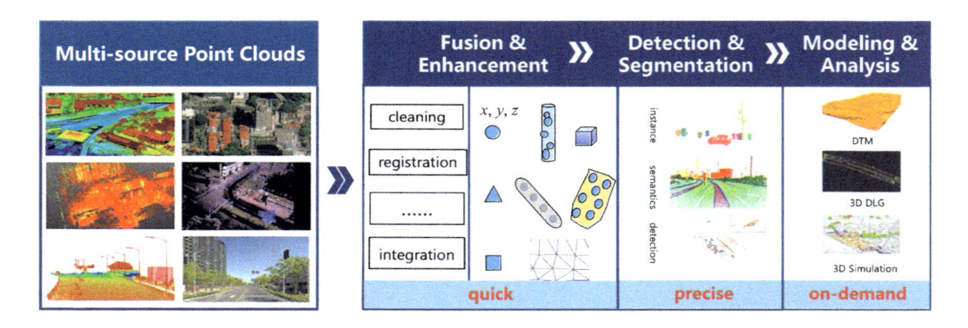

FIGURE 2.2 The framework of the ubiquitous point cloud theory.

category, and internal and external attributes. It abstractly represents a voxel (a volume element) with a specific volume and internal and external characteristics in space. It is the minimum unit for describing objects and geographical phenomena and the minimum computing unit for spatial computation and analysis. The ubiquitous point cloud theory fully realizes the complementary advantages between common point clouds (point clouds acquired from multiple platforms), transforming from manually-assisted classification to intelligent artificial scene understanding in processing models and transforming from visual and measurable-only to computation and analysis-supported expression models (Figure 2.2).

2.2 DATA FUSION AND ENHANCEMENT

Due to the limited observation range in a single viewpoint of a platform and the lack of uniform spatial reference among several observation platforms, it is necessary not only to fuse point clouds between stations or strips but also to fuse point clouds from multiple platforms (such as airborne, vehicle-mounted, ground-based stations, etc.), to obtain comprehensive spatial information of target areas. This compensates for the data gaps resulting from a single viewpoint and platform, achieving a complete and detailed digital representation of large-scale scenes (Dong et al. 2020; Wang and Solomon 2019). Furthermore, the fusion of laser point clouds and image data is required due to the limited representative ability with intensity information in representing target surfaces. This enables point clouds to possess high-precision three-dimensional geometric information and more abundant spectral information (Cui et al. 2022; Li et al. 2019). The fusion of multi-source data, such as point clouds from different stations/strips, point clouds from various platforms, and point clouds and images, requires the correlation of corresponding features. In response to the low efficiency and high cost of traditional manual registration methods, domestic and international researchers have explored statistical analysis methods based on geometric or texture feature correlations (W. Dai et al. 2019; Lahat, Adali, and Jutten 2015).

Due to the differences in imaging mechanisms, dimensions, scales, accuracies, and viewpoints among data from different platforms and sensors, there is still a lack of universally applicable methods for fusing and enhancing point clouds. To address this, Yang et al. (2016) developed a robust multi-structural feature (points, lines, surfaces) extraction method based on multi-view geometry and computer vision techniques for point clouds observed from multi-platforms. This method creates matching models for corresponding features, solving the challenges of automatic extraction and matching geometric features in weak intersections and small overlapping areas. They also established a consistency mapping model based on matching geometric features, developed an automated division strategy considering the accuracy differences and constraints of closed-loop construction in overlapping areas, and accurately estimated initial parameter values in the consistency mapping model. They further developed a global optimization solution method with division control and stepwise iteration to overcome the drawbacks of local convergence and global divergence in optimization, solving the non-linear optimization problems of model parameters, and achieving fully automatic unified positioning, orientation, and representation of multi-source and heterogeneous data within a large-scale 3D scene, as shown in Figure 2.3.

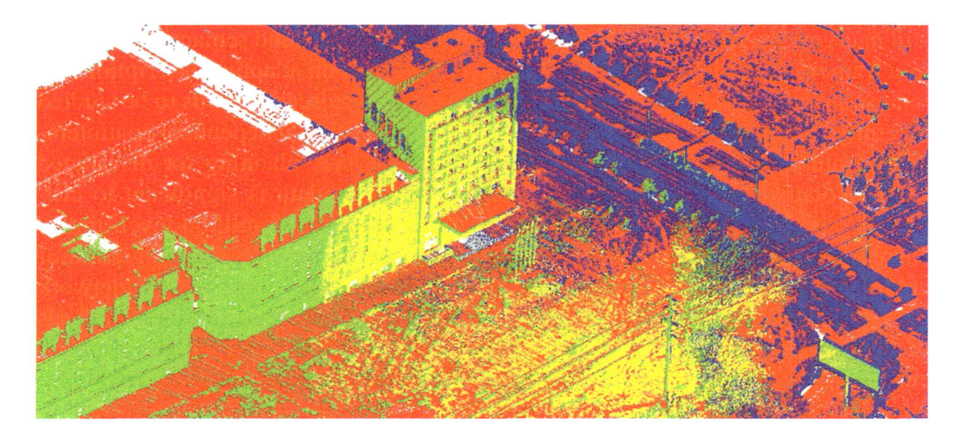

FIGURE 2.3 The illustration of the multi-source point clouds fusion in a large-scale 3D scenario.

2.3 OBJECT DETECTION AND SEGMENTATION

Object detection and segmentation are identifying and extracting artificial and natural elements from cluttered and unordered point clouds (Mi et al. 2022; Yang et al. 2015; Yang, Huang, et al. 2016; Yang, Dong, et al. 2017; Yang, Liu, et al. 2017). It serves as the foundation for subsequent applications such as digital terrain model generation and complex scenario reconstruction. However, source data from different platforms have different focuses regarding classifying laser point clouds. Aerial point cloud classification primarily focuses on targets such as large-scale ground, building roofs, vegetation, and roads (Yang, Huang, et al. 2016; Yang, Liu, et al. 2017). Vehicle-mounted point cloud classification, on the other hand, is concerned with targets including roads, road facilities on both sides, vegetation, and building facades (Yang, Dong, et al. 2017; Zhou et al. 2022). Ground-based point cloud classification emphasizes the fine interpretation of specific target areas (Z. Guo and Feng 2020). However, point cloud scenes exhibit diverse targets, complex morphological structures, target occlusion and overlap, and differences in spatial density, posing common challenges for the automatic detection and segmentation of 3D point clouds. For instance, point-wise feature-based point cloud classification methods (Gu, Wang, and Xie 2017; B. Guo et al. 2015), and segmentation and clustering-based point cloud classification methods (Yang, Dong, et al. 2017; Yang, Huang, et al. 2016) detect and extract targets from point clouds. However, due to inadequate feature description capabilities, object detection and segmentation quality may fail to meet application requirements, significantly limiting the utility of 3D geospatial information data. Currently, deep learning methods have effectively enhanced feature description and significantly improved performance in object detection and segmentation compared to traditional classification methods that overly rely on manually-defined features (Y. Guo et al. 2021). Although there is great potential in 2D scene classification and interpretation using deep learning techniques, numerous challenges remain unresolved regarding semantic and instance segmentation in 3D scenes (Han, Dong, and Yang 2021). Specifically, research on the semantic connotation, classification system, and encoding methods of all spatiotemporal elements, including buildings, streets, vegetation, and trees, is still in urgent demand to establish a spatial semantic model and semantic classification framework for urban areas. Furthermore, research on adaptive description and representation of elemental structures in complex urban scenes needs to be studied carefully to achieve multi-scale, multi-level, and position-independent spatiotemporal feature expression. Additionally, research on task planning for distributed data annotation with the assistance of crowdsourcing data is significant to efficiently construct high-level benchmark databases for multi-million-level, multi-source, and heterogeneous data of various target types through the integration of crowdsourced annotation information by a multi-granularity distributed network system, as shown in Figure 2.4. Lastly, research on automated extraction methods of geometric structure and semantic information for all elemental geographical entities using artificial intelligence techniques such as deep learning, reinforcement learning, and transfer learning is of great potential, aiming to enhance intelligent understanding of point clouds.

FIGURE 2.4 The sample library of point cloud panoramic annotation.

2.4 MODELING AND ANALYSIS

After object detection and point cloud segmentation in large-scale scenes, the target point clouds remain sparse, unordered, and highly redundant. They cannot explicitly represent the geometric structure of elements or spatial topology between targets, making it challenging to effectively meet the downstream application requirements. Therefore, the transformation of the discrete and unordered point clouds into a combination model of geometric primitives with topological relationships must be achieved through 3D scene representation. Two main categories of methods are commonly used: data-driven and model-driven (Xiong et al. 2015; Perera and Maas 2014). The major remaining problems and challenges include the automatic completion of 3D models to overcome the impact of local data gaps (Elberink and Vosselman 2009), automated and robust reconstruction of complex-shaped and structurally complicated objects, and the transition from visualization-oriented 3D models to computation and analysis-oriented 3D models to improve the usability and applicability of the results. Additionally, different applications have different requirements for the level of detail for various types of objects within the scene (Biljecki et al. 2014). Thus, the 3D scene representation needs to strengthen adaptive multi-level reconstruction methods for different 3D objects (Yang, Dong, et al. 2017; Biljecki, Ledoux, and Stoter 2016) and establish a multi-level representative model for scene-object-elements with correct semantics and structural mappings, as shown in Figure 2.5. Based on the constructed structural semantic model, menu-driven services can be provided for comprehensive governance in smart cities. For example, emergency management departments

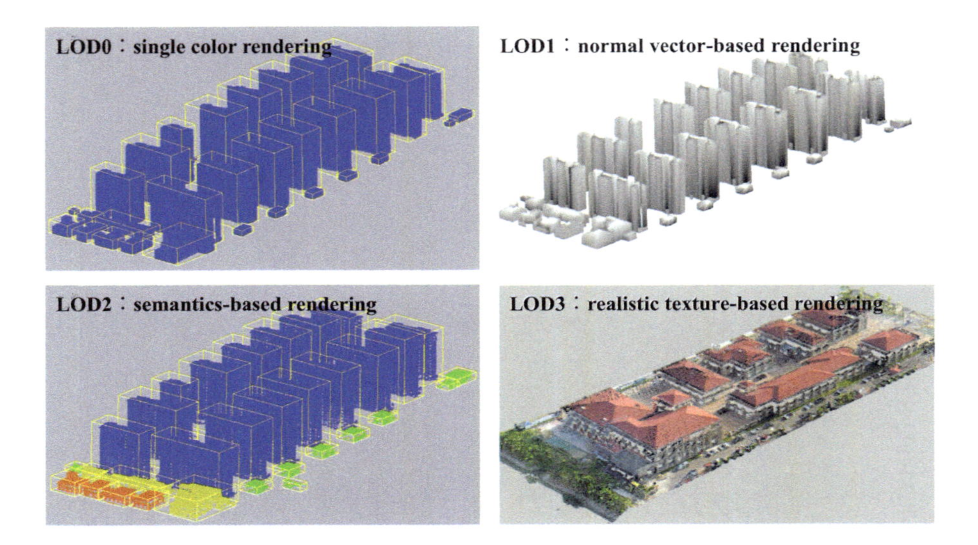

FIGURE 2.5 Task-driven modeling of urban elements.

require real-time and dynamic spatiotemporal data, water management departments need river data, green management departments need green space data, construction management departments need high-rise building data, and so on. These can be customized through intelligent filtering to extract the desired models. Currently, the main problems and challenges faced by data-driven and model-driven methods include: (1) exploring and constructing the regularity and correlation of geometry, semantics, attributes, and spatial relationships and dependencies between objects in complex scenarios, forming an element modeling grammar that includes "semantics-structure-relationships"; (2) an iterative and coupled approach based on "semantics-structure-relationships" for 3D semantic modeling of scenes, supporting structured semantic modeling and analysis of all elements, and forming an integrated and standardized expression method for constructing structured semantic models of scenes; (3) expanding the existing CityGML hierarchical model expression system (LoD) to form a multi-level expression system for urban elements that includes semantics, geometry, and texture information as needed.

Despite the significant progress made in data *fusion and enhancement*, object *detection and segmentation*, and *modeling and analysis* of ubiquitous point cloud theory, there is still a need for further breakthroughs in related research. On the one hand, there is a need to construct deep learning networks suitable for ultra-large-scale point cloud scenes. This involves direct learning from 3D point clouds, achieving end-to-end extraction of 3D objects, and structured reconstruction. Further research is required, such as loss function construction for 3D point cloud convolution (Song et al. 2018; Lang et al. 2019). On the other hand, although there are currently some open datasets available for 3D point clouds, such as NYU (Couprie et al. 2013), KITTI (Geiger, Lenz, and Urtasun 2012), ShapeNet (Chang et al. 2015), S3DIS (Armeni et al. 2017), ScanNet (A. Dai et al. 2017), Semantic 3D (Hackel et al. 2017), these datasets have limitations, such as a limited number of object categories, a small scene coverage, and inadequate scene diversity. Thus, there is an urgent need to construct more comprehensive, wide-ranging, and realistic benchmark datasets. Furthermore, the convergence of Internet of Things (IoT) data with point cloud digital infrastructure is needed. This involves developing techniques for optimizing spatiotemporal error coupling and multi-structure-constrained fusion of IoT data with point cloud spatial information. Key technologies include accurate matching of multi-modal data from IoT sensor data to point cloud coordinate system, enhancing the spatiotemporal consistency between the 3D digital foundation and IoT dynamic sensing data, such as video surveillance data, vehicle trajectory data, air quality data, hydrological data, meteorological data, and utility meter data. Besides, 5G, with ultra-high bandwidth, low latency, high reliability, comprehensive coverage, and massive connections, combined with edge computing capabilities, enables real-time transmission and online processing of point cloud big data. This will further promote the applications of point cloud big data in industries such as virtual/augmented reality, autonomous driving, power line inspection, and logistics delivery.

REFERENCES

Armeni, Iro, Sasha Sax, Amir R. Zamir, and Silvio Savarese. 2017. "Joint 2D-3D Semantic Data for Indoor Scene Understanding." *CoRR* abs/1702.01105. http://arxiv.org/abs/1702.01105

Biljecki, Filip, Hugo Ledoux, and Jantien Stoter. 2016. "An Improved LOD Specification for 3D Building Models." *Computers, Environment and Urban Systems* 59: 25–37. doi:10.1016/j.compenvurbsys.2016.04.005

Biljecki, Filip, Hugo Ledoux, Jantien Stoter, and Junqiao Zhao. 2014. "Formalisation of the Level of Detail in 3D City Modelling." *Computers, Environment and Urban Systems* 48: 1–15. doi:10.1016/j.compenvurbsys.2014.05.004

Chang, Angel X., Thomas A. Funkhouser, Leonidas J. Guibas, Pat Hanrahan, Qi-Xing Huang, Zimo Li, Silvio Savarese, et al. 2015. "ShapeNet: An Information-Rich 3D Model Repository." *CoRR* abs/1512.03012. http://arxiv.org/abs/1512.03012

Couprie, Camille, Clément Farabet, Laurent Najman, and Yann LeCun. 2013. "Indoor Semantic Segmentation Using Depth Information." *arXiv Preprint arXiv:1301.3572.*

Cui, Yaodong, Ren Chen, Wenbo Chu, Long Chen, Daxin Tian, Ying Li, and Dongpu Cao. 2022. "Deep Learning for Image and Point Cloud Fusion in Autonomous Driving: A Review." *IEEE Transactions on Intelligent Transportation Systems* 23 (2): 722–39. doi:10.1109/TITS.2020.3023541

Dai, Angela, Angel X. Chang, Manolis Savva, Maciej Halber, Thomas Funkhouser, and Matthias Niessner. 2017. "ScanNet: Richly-Annotated 3D Reconstructions of Indoor Scenes." *Proceedings of the IEEE Conference on Computer Vision and Pattern Recognition (CVPR)*, July.

Dai, Wenxia, Bisheng Yang, Xinlian Liang, Zhen Dong, Ronggang Huang, Yunsheng Wang, and Wuyan Li. 2019. "Automated Fusion of Forest Airborne and Terrestrial Point Clouds through Canopy Density Analysis." *ISPRS Journal of Photogrammetry and Remote Sensing* 156: 94–107. doi:10.1016/j.isprsjprs.2019.08.008

Dong, Zhen, Fuxun Liang, Bisheng Yang, Yusheng Xu, Yufu Zang, Jianping Li, Yuan Wang, et al. 2020. "Registration of Large-Scale Terrestrial Laser Scanner Point Clouds: A Review and Benchmark." *ISPRS Journal of Photogrammetry and Remote Sensing* 163: 327–42. doi:10.1016/j.isprsjprs.2020.03.013

Elberink, Sander Oude, and George Vosselman. 2009. "Building Reconstruction by Target Based Graph Matching on Incomplete Laser Data: Analysis and Limitations." *Sensors* 9 (8): 6101–18. doi:10.3390/s90806101

Geiger, Andreas, Philip Lenz, and Raquel Urtasun. 2012. "Are We Ready for Autonomous Driving The KITTI Vision Benchmark Suite." In *2012 IEEE Conference on Computer Vision and Pattern Recognition*, 3354–61. doi:10.1109/CVPR.2012.6248074

Gu, Yanfeng, Qingwang Wang, and Bingqian Xie. 2017. "Multiple Kernel Sparse Representation for Airborne LiDAR Data Classification." *IEEE Transactions on Geoscience and Remote Sensing* 55 (2): 1085–1105. doi:10.1109/TGRS.2016.2619384

Guo, Bo, Xianfeng Huang, Fan Zhang, and Gunho Sohn. 2015. "Classification of Airborne Laser Scanning Data Using JointBoost." *ISPRS Journal of Photogrammetry and Remote Sensing* 100. 71–83.

Guo, Yulan, Hanyun Wang, Qingyong Hu, Hao Liu, Li Liu, and Mohammed Bennamoun. 2021. "Deep Learning for 3D Point Clouds: A Survey." *IEEE Transactions on Pattern Analysis and Machine Intelligence* 43 (12): 4338–64. doi:10.1109/TPAMI.2020.3005434

Guo, Zhou, and Chen-Chieh Feng. 2020. "Using Multi-Scale and Hierarchical Deep Convolutional Features for 3D Semantic Classification of TLS Point Clouds." *International Journal of Geographical Information Science* 34 (4). 661–80. doi:10.1080/13658816.2018.1552790

Hackel, Timo, Nikolay Savinov, Lubor Ladicky, Jan Dirk Wegner, Konrad Schindler, and Marc Pollefeys. 2017. "Semantic3D Net: A New Large-Scale Point Cloud Classification Benchmark." *CoRR* abs/1704.03847. http://arxiv.org/abs/1704.03847

Han, Xu, Zhen Dong, and Bisheng Yang. 2021. "A Point-Based Deep Learning Network for Semantic Segmentation of MLS Point Clouds." *ISPRS Journal of Photogrammetry and Remote Sensing* 175: 199–214. https://doi.org/10.1016/j.isprsjprs.2021.03.001

Lahat, Dana, Tülay Adali, and Christian Jutten. 2015. "Multimodal Data Fusion: An Overview of Methods, Challenges, and Prospects." *Proceedings of the IEEE* 103 (9): 1449–77. doi:10.1109/JPROC.2015.2460697

Lang, Alex H., Sourabh Vora, Holger Caesar, Lubing Zhou, Jiong Yang, and Oscar Beijbom. 2019. "PointPillars: Fast Encoders for Object Detection from Point Clouds." In *Proceedings of the IEEE/CVF Conference on Computer Vision and Pattern Recognition (CVPR).*

Li, Jianping, Bisheng Yang, Chi Chen, and Ayman Habib. 2019. "NRLI-UAV: Non-Rigid Registration of Sequential Raw Laser Scans and Images for Low-Cost UAV LiDAR Point Cloud Quality Improvement." *ISPRS Journal of Photogrammetry and Remote Sensing* 158: 123–45. doi:10.1016/j.isprsjprs.2019.10.009

Mi, Xiaoxin, Bisheng Yang, Zhen Dong, Chi Chen, and Jianxiang Gu. 2022. "Automated 3D Road Boundary Extraction and Vectorization Using MLS Point Clouds." *IEEE Transactions on Intelligent Transportation Systems* 23 (6): 5287–97. doi:10.1109/TITS.2021.3052882

Perera, Gamage Sanka Nirodha, and Hans-Gerd Maas. 2014. "Cycle Graph Analysis for 3D Roof Structure Modelling: Concepts and Performance." *ISPRS Journal of Photogrammetry and Remote Sensing* 93: 213–26. doi:10.1016/j.isprsjprs.2014.04.017

Song, Yuhang, Chao Yang, Yeji Shen, Peng Wang, Qin Huang, and C.-C. Jay Kuo. 2018. "SPG-Net: Segmentation Prediction and Guidance Network for Image Inpainting." *CoRR* abs/1805.03356. http://arxiv.org/abs/1805.03356

Wang, Yue, and Justin M. Solomon. 2019. "Deep Closest Point: Learning Representations for Point Cloud Registration." In *Proceedings of the IEEE/CVF International Conference on Computer Vision (ICCV).*

Xiong, B., M. Jancosek, S. Oude Elberink, and G. Vosselman. 2015. "Flexible Building Primitives for 3D Building Modeling." *ISPRS Journal of Photogrammetry and Remote Sensing* 101: 275–90. doi:10.1016/j.isprsjprs.2015.01.002

Yang, Bisheng, Zhen Dong, Fuxun Liang, and Yuan Liu. 2016. "Automatic Registration of Large-Scale Urban Scene Point Clouds Based on Semantic Feature Points." *ISPRS Journal of Photogrammetry and Remote Sensing* 113: 43–58. doi:10.1016/j.isprsjprs.2015.12.005

Yang, Bisheng, Zhen Dong, Yuan Liu, Fuxun Liang, and Yongjun Wang. 2017. "Computing Multiple Aggregation Levels and Contextual Features for Road Facilities Recognition Using Mobile Laser Scanning Data." *ISPRS Journal of Photogrammetry and Remote Sensing* 126: 180–94. doi:10.1016/j.isprsjprs.2017.02.014

Yang, Bisheng, Zhen Dong, Gang Zhao, and Wenxia Dai. 2015. "Hierarchical Extraction of Urban Objects from Mobile Laser Scanning Data." *ISPRS Journal of Photogrammetry and Remote Sensing* 99: 45–57. doi:10.1016/j.isprsjprs.2014.10.005

Yang, Bisheng, Ronggang Huang, Zhen Dong, Yufu Zang, and Jianping Li. 2016. "Two-Step Adaptive Extraction Method for Ground Points and Breaklines from Lidar Point Clouds." *ISPRS Journal of Photogrammetry and Remote Sensing* 119: 373–89. doi:10.1016/j.isprsjprs.2016.07.002

Yang, Bisheng, Yuan Liu, Zhen Dong, Fuxun Liang, Bijun Li, and Xiangyang Peng. 2017. "3D Local Feature BKD to Extract Road Information from Mobile Laser Scanning Point Clouds." *ISPRS Journal of Photogrammetry and Remote Sensing* 130: 329–43. doi:10.1016/j.isprsjprs.2017.06.007

Yang, Bisheng, and Jinling Wang. 2016. "Mobile Mapping with Ubiquitous Point Clouds." *Geo-Spatial Information Science* 19 (3). 169–70. doi:10.1080/10095020.2016.1244982

Zhou, Yuzhou, Xu Han, Mingjun Peng, Haiting Li, Bo Yang, Zhen Dong, and Bisheng Yang. 2022. "Street-View Imagery Guided Street Furniture Inventory from Mobile Laser Scanning Point Clouds." *ISPRS Journal of Photogrammetry and Remote Sensing* 189: 63–77. doi:10.1016/j.isprsjprs.2022.04.023

Part II

Fusion and Enhancement of Point Cloud

There are currently many acquisition platforms for point cloud, including Airborne Laser Scanning (ALS), Mobile Laser Scanning (MLS), Terrestrial Laser Scanning (TLS), etc. However, limited to the scanning perspective and resolution, it is difficult to fully describe the three-dimensional scene using point clouds from only one platform. It is essential to fuse and enhance point clouds to meet applications in more scenarios. Recent research works have achieved good performance on pairwise registration, whereas the fusion of point cloud in terms of multi-view, cross-platform, and cross-modal is still challenging.

This part first introduces a method for the multi-view registration of point clouds. Previous multi-view registration methods commonly rely on exhaustive pairwise registration to construct a densely connected pose graph and apply Iteratively Reweighted Least Square (IRLS) on the pose graph to compute the scan poses. The construction is time-consuming and contains many outlier edges, making the subsequent IRLS struggle to find correct poses. To address the above problems, we use a neural network to estimate the overlap between scan pairs, which enables us to construct a sparse but reliable pose graph. Then, a novel history reweighting function is adopted in the IRLS scheme, which has strong robustness to outlier edges on the graph. Compared with other multi-view registration methods, the method achieves higher registration recall and lower registration errors while reducing required pairwise registrations. Comprehensive ablation studies are conducted to demonstrate the effectiveness.

Additionally, a method of registering TLS to ALS is introduced for cross-platform registration. The automatic registration of multiple TLS scans to ALS is challenging due to the different perspective views, resolutions, and ranges. A skyline context-based method is presented to address this issue. First, the ground points in ALS are extracted and used as potential TLS locations, and the corresponding skyline contexts are generated. After that, a 3D skyline-based k-d tree is built to search the corresponding coarse TLS scan localizations. The trimmed iterative closest point (T-ICP) algorithm completes the final refinement. Five datasets with different ALS sizes and over one hundred TLS scans are undertaken to evaluate the performance of the proposed method. Experimental results indicate that the proposed method performs well for automatic fusion of TLS scans to ALS point clouds, with advantages in both precision and adaptability.

Regarding cross-modality, we use a high-precision registration method of point clouds and images to generate colored point clouds with texture attributes, which improves point cloud classification and target extraction results. Although existing commercial drone systems and vehicle-mounted systems have provided system calibration parameters, due to the differences in the dimensions and sampling granularity and the complex mapping relationship between the point clouds and images, it is still difficult to establish an accurate correspondence between them. To tackle this issue, we proposed a novel automatic method to register the panoramic images and MLS point clouds based on matching pole objects. 2D and 3D pole objects are extracted and re-projected onto virtual images, respectively, and then fine 2D-3D correspondences are collected through maximizing pole overlap by Particle Swarm Optimization (PSO). Experiments indicate that the proposed method performs effectively on two challenging urban scenes.

In general, this part gives several validated methods in dealing with multi-view, cross-platform, and cross-modal fusion, providing a broader perspective for the fusion of point clouds. Benefiting from these methods, the capability of point cloud data in capturing and representing the three-dimensional world has experienced remarkable expansion, which is quite conducive to subsequent applications.

3 Multiview Point Clouds Registration

3.1 METHODOLOGY

3.1.1 Overview

Consider a set of unaligned scans $\mathcal{P} = \{P_i \mid i = 1,\dots N\}$ in the same 3D scene. The target of multi-view registration is to recover the underlying global scan poses $\{T_i = (R_i, t_i) \in SE(3) \mid i = 1, \dots, N\}$. Previous multi-view registration methods rely on exhaustive pairwise registration to construct a densely-connected pose graph and apply Iteratively Reweighted Least Square (IRLS) on the pose graph to compute the scan poses. However, constructing a densely-connected graph is time-consuming and contains lots of outlier edges, which makes the subsequent IRLS struggle to find correct poses. As shown in Figure 3.1, to address the above problems, we first propose to use a neural network to estimate the overlap between scan pairs, which enables us to construct a *sparse but reliable* pose graph. Then, we design a novel *history reweighting function* in the IRLS scheme, which has strong robustness to outlier edges on the graph.

3.1.2 Learn to Construct a Sparse Graph

In this section, we aim to construct a pose graph for the multi-view registration. Specifically, the graph is denoted by $\mathcal{G}(\mathcal{V},\mathcal{E})$, where each vertex $v_i \in \mathcal{V}$ represents each scan P_i while edge $(i,j) \in \mathcal{E}$ encodes the relative poses between scan P_j and scan P_i. We will first estimate an overlap score s_{ij} for each scan pair (P_i, P_j). Then, given the overlap scores, we construct a sparse graph by selecting a set of scan pairs with large estimated overlaps and apply pairwise transformations on them only.

Global feature extraction. To extract the global feature F for a point cloud P, we first downsample P by voxels and extract a local feature $f_p \in R^d$ on every sampled point $p \in P$ from its local 3D patch $N_p = p' \mid \|p - p'\|_2 < r, p' \in P$ within a radius of r by $f_p = \varphi(N_p)$, where φ is a neural network module for extracting local descriptors, specifically YOHO (Wang et al. 2022) due to its superior performance. Then, we apply a *NetVLAD* (Arandjelovic et al. 2016)

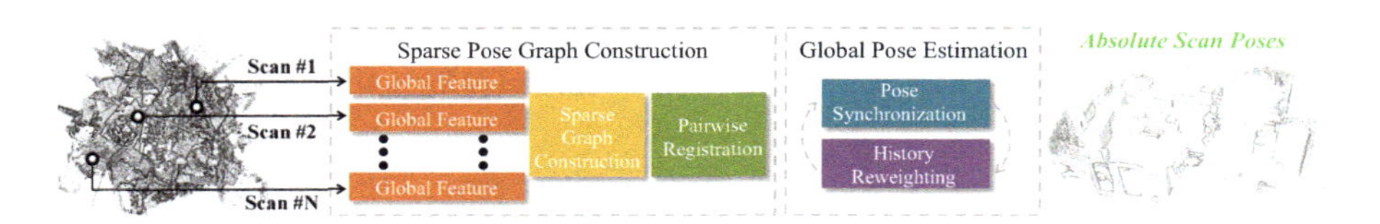

FIGURE 3.1 Overview of the proposed multi-view registration method.

DOI: 10.1201/9781003486060-5

layer on the local features to extract a global feature $F = NetVLAD(\{f_p\})$. Note $F \in R^n$ is normalized such that $||F||_2 = 1$.

Sparse graph construction. For a scan pair (P_i, P_j), we estimate their overlap score by $s_{ij} = \left(F_i^T F_j + 1\right)/2$, where $s_{ij} \in [0, 1]$ indicates the overlap between P_i and P_j. We train the *NetVLAD* with a l_1 loss between the predicted overlap score and the ground-truth overlap ratio. For each scan, we select other k scan pairs with the largest overlap scores to connect with the scan. This leads to a sparse graph with edges $\mathcal{E} = \left\{\left(\mathrm{i}, \mathrm{j} : arg\ topk_{P_j \in \mathcal{P}, j \neq i}\ \mathrm{s_{ij}}\right), \forall \mathrm{P_i} \in \mathcal{P}\right\}$. On each edge $(i, j) \in \mathcal{E}$ of the constructed graph, we estimate a relative pose T_{ij} on the scan pair from their extracted local descriptors. By default, we follow Wang et al. (2022) to apply the nearest neighborhood matcher on the local descriptors and estimate the relative pose from the RANSAC variant.

Discussion. Recent multi-view registration methods (Dong et al. 2017; Huang et al. 2021; Wang et al. 2022; Qin et al. 2023) usually exhaustively estimate all $\binom{N}{2}$ relative poses and many of these scan pairs have no overlap at all. In our method, we extract global features to determine the overlap scores to select $N \times k$ scan pairs. Actually, we only need to conduct pairwise registration less than $N \times k$ times because the graph is an indirection graph and each edge counts once only. Our global feature extraction is much more efficient than matching descriptors and running RANSAC in pairwise registration.

The subsequent pose synchronization only considers these sparse edges, which also improves the efficiency. Moreover, the retained pose graph contains much fewer outliers, which thus improves the accuracy of the subsequent synchronization.

3.1.3 IRLS with History Reweighting

In this section, we apply the Iteratively Reweighted Least Squares (IRLS) scheme to estimate the consistent global poses on all scans. The key idea of IRLS is to associate a weight on each edge to indicate the reliability of each scan pair. These weights are iteratively refined such that outlier edges will have small weights so these outlier relative poses will not affect the final global poses. In the following, we first initialize edge weights, and iteratively estimate poses based on edge weights and update edge weights with the proposed history reweighting function.

3.1.3.1 Weight Initialization

The weight $w_{ij}^{(0)}$ is initialized from both the estimated overlap score s_{ij} and the quality of the pairwise registration by $w_{ij}^{(0)} = s_{ij} * r_{ij}$, where r_{ij} reveals the quality of pairwise registration. In the pairwise registration, a set of correspondences $C = \{(p, q) | p \in P_i, q \in P_j\}$ are established by matching local descriptors. Thus, r_{ij} is defined as the number of inlier correspondences in C conforming with $T_{ij} = (R_{ij}, t_{ij})$, which is

$$r_{ij} = \sum_{(p,q) \in C} [\![\| p - R_{ij} q - t_{ij} \|_2 < \tau]\!], \tag{3.1}$$

where $[\![.]\!]$ is the Iverson bracket, τ is a pre-defined inlier threshold.

3.1.3.2 Pose Synchronization

Given the edge weights and input relative poses $w_{ij}, T_{ij} = (R_{ij}, t_{ij}) | (i,j) \in \mathcal{E}$, we solve for the global scan poses $\{T_i = (R_i, t_i)\}$. We adopt the closed-form synchronization algorithm (Gojcic et al. 2020; Yew et al. 2021). We first compute the rotations by rotation synchronization, and then compute the translations by translation synchronization.

Rotation synchronization. The goal of rotation synchronization is to solve

$$\{R_1, \dots R_N\} = argmin_{R_1,\dots R_N \in SO(3)} \sum_{(i,j)\in\mathcal{E}} w_{ij} \| R_{ij} - R_i^T R_j \|_F^2 \tag{3.2}$$

where $\| \cdot \|_F$ means the Frobenius norm of the matrix.

The problem has a closed-form solution, which can be derived from the eigenvectors of a symmetric matrix $L \in \mathbb{R}^{3N*3N}$:

$$L = \begin{pmatrix} \sum_{(1,j)\in\mathcal{E}} w_{1j}\mathbf{I_3} & -w_{12}R_{12} & \cdots & -w_{1N}R_{1N} \\ -w_{21}R_{21} & \sum_{(2,j)\in\mathcal{E}} w_{2j}\mathbf{I_3} & \cdots & -w_{2N}R_{2N} \\ \vdots & \vdots & \ddots & \vdots \\ -w_{N1}R_{N1} & -w_{N2}R_{N2} & \cdots & \sum_{(N,j)\in\mathcal{E}} w_{Nj}\mathbf{I_3} \end{pmatrix}, \tag{3.3}$$

where $I_3 \in \mathbb{R}^{3*3}$ denotes the identity matrix. L is a sparse matrix since the constructed graph is sparse and the corresponding blocks of unconnected edges are set to zeros. Given three eigenvectors $\tau_1, \tau_2, \tau_3 \in \mathbb{R}^{3*3N}$ corresponding to the three smallest eigenvalues $\lambda_1 < \lambda_2 < \lambda_3$ of L, we stack these three eigenvectors to construct a matrix $\gamma = [\tau_1, \tau_2, \tau_3] \in \mathbb{R}^{3N*3}$. Then, $v_i = \gamma[3i - 3 : 3i] \in \mathbb{R}^{3*3}$ is an approximation of the absolute rotation R_i for point cloud P_i but may not satisfy the constraint $v_i v_i^T = \mathbf{I}_3$. Therefore, we rectify this by applying singular value decomposition on v_i by $v_i = U_i \sum_i V_i^T$ and deriving $R_i = V_i U_i^T$ (Arie-Nachimson et al. 2012). Then, we further check $det(R_i)$ and exchange the first two rows of R_i if $det(R_i) = -1$.

Translation synchronization. Translation synchronization retrieves the translation vectors $\{t_i\}$ that minimize the problem:

$$\{t_1, \dots t_N\} = argmin_{t_1,\dots t_N \in \mathbb{R}^3} \sum_{(i,j)\in\mathcal{E}} w_{ij} \| R_i t_{ij} - t_j + t_i \|_2 . \tag{3.4}$$

We solve it by the standard least square method (Huang et al. 2017). Assuming E edges are connected in $\mathcal{G}$, we thus construct three matrices A, B, and H as follows. $A \in \mathbb{R}^{3E*3E}$ is initialized as an identity matrix. $B \in \mathbb{R}^{3E*3N}$ contains $E * N$ 3×3 blocks and is initialized as a zero matrix. $H \in \mathbb{R}^{3E*1}$ is a vector containing E 3×1 blocks.

For the *e-th* edge $(i,j) \in \mathcal{E}$, we multiply $A[3e - 3 : 3e]$ with w_{ij}, fill $\boldsymbol{I_3}$ and $-\boldsymbol{I_3}$ to the (e, j) and (e, i) block of B respectively, and fill $R_i t_{ij}$ to the *e-th* block of H. We thus solve $t = (B^T AB)^{-1} B^T AH$ and obtain the translation vector t_i of each scan P_i as $t[3i - 3:3i]$.

History reweighting function. Given the synchronized poses, we recompute weights on edges such that outlier edges will have smaller weights than the inlier edges. Assume the synchronized poses at the n-th iteration are $\left\{T_i^{(n)} = \left(R_i^{(n)}, t_i^{(n)}\right)\right\}$. We first compute the rotation residual $\delta_{ij}^{(n)}$ by

$$\delta_{ij}^{(n)} = \Delta\left(R_{ij}, R_i^{(n)T} R_j^{(n)}\right), \tag{3.5}$$

where $\Delta(R_1, R_2)$ means the angular difference between the rotation R_1 and R_2. $\Delta(R_1, R_2)$ is implemented by transforming $R_1^T R_2$ into an axis-angle form and outputing the rotation angle. Then, the updated weights are computed from rotation residuals of all previous iterations by

$$w_{ij}^{(n)} = w_{ij}^{(0)} exp\left(\sum_{m=1}^{n} g(m)\delta_{i,j}^{(m)}\right), \tag{3.6}$$

where $g(m)$ is a pre-defined coefficient function of the iteration number with $g(m) > 0$. We first discuss the intuition behind the Eq. (3.6), and then elaborate the design of $g(m)$.

Intuition of Eq. (3.6). Similar to previous reweighting functions (Arrigoni et al. 2016, Gojcic et al. 2020; Yew et al. 2021), a larger rotation residual δ will lead to a smaller weight because large residuals are often caused by outliers. Meanwhile, there are two differences from previous reweighting functions. First, we multiply the initial weights $w_{ij}^{(0)}$ so that the recomputed weights always retain information from the warm-start initialization and these initialized weights will be adjusted by the residuals in the iterative refinement. Second, the weight at a specific iteration n considers the residuals of all previous iterations $m \leq n$. This design is inspired by the momentum optimization method RMSProp or Adam (Kingma and Ba, 2014), which utilizes the gradients in the history to stabilize the optimization process. Here, we adopt a similar strategy to consider all residuals in the history to determine a robust weight for the current iteration, which is less sensitive to outliers.

Design of coefficient function $g(m)$. $g(m)$ can be regarded as a weight function. A small value of $g(m)$ means that we do not trust the residual at the iteration m and this residual may not correctly identify inliers and outliers. In our observation, the residuals estimated by the first few iterations are not very stable so the $g(m)$ is increasing with the iteration number m. Meanwhile, if we want to conduct M IRLS iterations in total, the sum of coefficients at the final iteration M should be 1, i.e. $\sum_{m=1}^{M} g(m) = 1$. Thus, in our design, we have

$$g(m) = \frac{2m}{M(M+1)}. \tag{3.7}$$

After computing the updated weights, we iteratively synchronize the poses with these updated weights and compute new weights from these new poses with transformation synchronization. The IRLS run M iterations in total and the synchronized poses at the final iteration are regarded as the output poses for all scans.

3.2 EXPERIMENTS

3.2.1 Experimental Protocol

3.2.1.1 Datasets

We evaluate the proposed method on three widely used datasets.

3DMatch contains scans collected from 62 indoor scenes among which 46 are split for training, eight for validation, and eight for testing. Each test scene contains 54 scans on average. We follow previous works to use 1623 scan pairs with >30% overlap ratio and 1781 scan pairs with 10% ~ 30% overlap as two test sets, denoted as 3DMatch (Zeng et al. 2017) and 3DLoMatch (Huang et al. 2021), respectively.

ScanNet contains RGBD sequences of 1513 indoor scenes. We follow Wang et al. (2022) to use the same 32 test scenes and convert 30 RGBD images that are 20 frames apart to 30 scans on each scene. There are 960 scans in total and we exhaustively select all 13,920 scan pairs for evaluation.

ETH has four outdoor scenes with large domain gaps to indoor datasets and each scene contains 33 scans. 713 scan pairs are officially selected for evaluation (Pomerleau et al. 2012).

Our model is only trained on the training split of 3DMatch and evaluated on 3D(Lo)Match, ScanNet, and ETH. For evaluation, we first perform multi-view registration to recover the global scan poses. Then, we follow Gojcic et al. (2020) to evaluate the multi-view registration quality on pairwise relative poses computed from the recovered global poses. By default, we set k in sparse graph construction to ten for two indoor datasets and six for the ETH dataset.

3.2.1.2 Metrics

We follow (Huang et al. 2021; Wang et al. 2022; Qin et al. 2023) to adopt Registration Recall (RR) for evaluation on 3D(Lo)Match and ETH. RR reports the ratio of correctly-aligned scan pairs. A scan pair is regarded as correctly-aligned if the average distance between the points under the estimated transformation (R_{pre}, t_{pre}) and these points under the ground truth transformation (R_{gt}, t_{gt}) is less than 0.2 m for the 3D(Lo)Match dataset and 0.5 m for the ETH dataset. RR of all methods is calculated on the same official evaluation scan pairs. For the evaluation on ScanNet, we follow (Gojcic et al. 2020; Yew et al. 2021) to report Empirical Cumulative Distribution Functions (ECDF) of the rotation error re and translation error te:

$$re = arccos\left(\frac{tr\left(R_{pre}^{T}R_{gt}\right)-1}{2}\right), te = \left\| t_{pre} - t_{gt} \right\|_2. \tag{3.8}$$

We also report the number of required pairwise registrations to initialize the pose graphs, denoted as "#Pair".

3.2.1.3 Baselines

We compare the proposed method against several multi-view registration baselines: EIGSE3 (Arrigoni et al. 2016), L1-IRLS (Chatterjee and Govindu, 2017), RotAvg (Chatterjee and Govindu, 2017), LMVR (Gojcic et al. 2020), LITS (Yew et al. 2021), and HARA (Lee and Civera, 2022). Specifically, LMVR is an end-to-end method, which performs pairwise registration and transformation synchronization in a single deep neural network. EIGSE3 proposes a spectral approach to solve transformation synchronization and further applies IRLS with the Cauchy reweighting function to improve robustness. L1-IRLS and RotAvg are two robust algorithms, which perform IRLS-based rotation synchronization using l_1 and $l_{1/2}$ reweighting functions to resist outliers.

HARA is a state-of-the-art handcrafted synchronization method, which conducts growing-based edge pruning by checking cycle consistency and performs IRLS-based synchronization using $l_{1/2}$ reweighting functions on the retained edges. LITS is a state-of-the-art learning-based transformation synchronization method. All baseline methods except LMVR are compatible with any pairwise registration methods. Thus, we compare our method with these baselines using different pairwise registration algorithms, including FCGF (Choy et al. 2019), SpinNet (Ao et al. 2021), YOHO (Wang et al. 2022), GeoTransformer (Qin et al. 2023).

3.2.1.4 Pose Graph Construction

For a fair comparison with baseline multi-view registration methods, we report the performances produced on three different types of input pose graphs. The first type "Full" does not prune any edge so the pose graph is fully connected. The second type "Pruned" prunes edges according to the quality of pairwise registration, which is adopted by previous methods LMVR (Gojcic et al. 2020) and LITS (Yew et al. 2021) (called "Good" in their papers). "Pruned" first applies pairwise registration algorithms to exhaustively register all scan pairs and then only retain scan pairs whose median point distance in the registered overlapping region is less than 0.05 m (0.15 m for ETH).

The final type "Ours" applies the proposed global feature for the overlap score estimation and constructs a sparse graph according to scores.

3.2.2 Results on Three Benchmarks

Results on the 3DMatch, the ScanNet and the ETH datasets are shown in Tables 3.1, 3.2 and 3.3, respectively.

First, the results show that our method achieves significant better performances than all baseline methods with 5%–10% improvements on the 3DMatch and 3DLoMatch dataset, which demonstrates that our method is able to accurately align low-overlapped scan pairs via pose synchronization. Meanwhile, our method only requires 30% pairwise registrations with the help of our sparse graph construction, which greatly improves the efficiency.

Second, when using the same pose graphs as previous method, our method already achieves better performances on all datasets, which is benefited from our history

TABLE 3.1

Registration Recall on the 3DMatch ("3D") and 3DLoMatch ("3DL") Datasets

Pose Graph	Method	#Pair	SpinNet	YOHO	GeoTrans
			3D / 3DL-RR (%)	3D / 3DL-RR (%)	3D / 3DL-RR (%)
Full	EIGSE3	11905	20.8/13.6	23.2/6.6	17.0/9.1
	L1-IRLS	11905	49.8/29.4	52.2/32.2	55.7/37.3
	RotAvg	11905	59.3/38.9	61.8/44.1	68.6/56.5
	LITS	11905	68.1/47.9	77.0/59.0	84.2/73.0
	HARA	11905	82.7/63.6	83.1/68.7	83.4/68.5
	Ours	11905	93.3/77.2	93.2/76.8	91.5/82.4
Pruned	EIGSE3	11905	42.7/34.6	40.1/26.5	39.4/28.7
	L1-IRLS	11905	66.9/46.2	68.6/49.0	77.4/58.3
	RotAvg	11905	72.8/55.3	77.2/60.3	81.6/68.5
	LITS	11905	73.1/55.5	80.8/65.2	84.6/76.8
	HARA	11905	84.0/62.5	83.8/71.9	84.9/73.7
	Ours	11905	94.8/**80.6**	95.2/**82.3**	95.2/82.8
Ours	Ours	**2798**	**94.9**/80.0	**96.2**/81.6	**95.9/83.0**

Note: We report results with different pairwise registration algorithms (SpinNet, YOHO, GeoTrans).

TABLE 3.2
Registration Performance on the ScanNet Dataset

Pose Graph	Method	#Pair	Rotation Error						Translation Error (m)					
			3°	5°	10°	30°	45°	Mean/Med	0.05	0.1	0.25	0.5	0.75	Mean/Med
Full	LMVR	13920	48.3	53.6	58.9	63.2	64.0	48.1°/33.7°	34.5	49.1	58.5	61.6	63.9	0.83/0.55
	LITS	13920	47.4	58.4	70.5	78.3	79.7	27.6°/-	29.6	47.5	66.7	73.3	77.6	0.56/-
	EIGSE3*	13920	19.7	24.4	32.3	49.3	56.9	53.6°/48.0°	11.2	19.7	30.5	45.7	56.7	1.03/0.94
	L1-IRLS*	13920	38.1	44.2	48.8	55.7	56.5	53.9°/47.1°	18.5	30.4	40.7	47.8	54.4	1.14/1.07
	RotAvg*	13920	44.1	49.8	52.8	56.5	57.3	53.1°/44.0°	28.2	40.8	48.6	51.9	56.1	1.13/1.05
	LITS*	13920	52.8	67.1	74.9	77.9	79.5	26.8°/27.9°	29.4	51.1	68.9	75.0	77.0	0.68/0.66
	HARA*	13920	54.9	64.3	71.3	74.1	74.2	32.1°/29.2°	35.8	54.4	66.3	69.7	72.9	0.87/0.75
	Ours	13920	57.2	68.5	75.1	78.1	78.8	26.4°/19.5°	39.4	61.5	72.0	75.2	77.6	0.70/0.59
Pruned	EIGSE3*	13920	40.8	46.3	51.9	61.2	65.7	40.6°/37.1°	23.9	38.5	51.0	59.3	66.1	0.88/0.84
	L1-IRLS*	13920	46.3	54.2	61.6	64.3	66.8	41.8°/34.0°	24.1	38.5	48.3	55.6	60.9	1.05/1.01
	RotAvg*	13920	50.2	60.1	65.3	66.8	68.8	38.5°/31.6°	31.8	49.0	58.8	63.3	65.6	0.96/0.83
	LITS*	13920	54.3	69.4	75.6	78.5	80.3	24.9°/19.9°	31.4	54.4	72.3	76.7	79.6	0.65/0.56
	HARA*	13920	55.7	63.7	69.0	70.8	72.1	34.7°/31.3°	35.2	53.6	65.4	68.6	71.7	0.86/0.71
	Ours	13920	**59.4**	71.9	80.0	82.1	82.6	**21.7**°/19.1°	**39.9**	63.0	74.3	77.6	80.2	0.64/**0.47**
Ours	Ours	**6004**	59.1	**73.1**	**80.8**	**82.5**	**83.0**	**21.7°/19.0°**	**39.9**	**64.1**	**76.7**	**79.0**	**81.9**	**0.56**/0.49

Note: The pairwise registration algorithm for all methods is YOHO except for LMVR.

* Means using the same selected frames and pairwise transformations as ours

TABLE 3.3
Registration Recall on the ETH Dataset

Pose Graph	Method	#Pair	FCGF RR (%)	SpinNet RR (%)	YOHO RR (%)
Full	EIGSE3	2123	44.8	56.3	60.9
	L1-IRLS	2123	60.5	73.2	77.2
	RotAvg	2123	67.3	82.1	85.4
	LITS	2123	26.3	36.4	34.8
	HARA	2123	72.2	79.3	85.4
	Ours	2123	85.7	86.3	98.8
Pruned	EIGSE3	2123	89.4	93.6	96.3
	L1-IRLS	2123	86.1	87.9	90.2
	RotAvg	2123	95.6	95.5	96.6
	LITS	2123	41.2	47.3	48.4
	HARA	2123	90.3	97.8	96.0
	Ours	2123	96.8	**99.8**	97.2
Ours	Ours	**516**	**97.4**	**99.8**	**99.1**

Note: We report results using different pairwise registration algorithms

reweighting function in the IRLS. Meanwhile, applying our global features for the graph construction further improves the results, which demonstrates the predicted overlap score is more robust than simply pruning edges according to the pairwise registration.

Finally, the results on the outdoor ETH dataset demonstrate the generalization ability of the proposed method. Both our method and the learning-based method LITS (Yew et al. 2021) are trained on the indoor 3DMatch dataset, However, LITS does not generalize well to outdoor datasets (only 45% recall) even though it shows strong performances on both indoor datasets. In comparison, our method still achieves strong performances (almost 100% registration recall) on the outdoor dataset.

3.2.3 Qualitative Results

Qualitative results on the three benchmarks are shown in Figure 3.2. We also provide the registration results of the proposed method on the large-scale WHU-TLS dataset (Dong et al. 2020) to show its effectiveness on various scenarios.

3.3 ANALYSIS AND DISCUSSION

We thoroughly conduct analyses on the proposed designs about the pose graph construction and history reweighting IRLS modules, the time-consuming analysis, and limitation analysis in this section. By default, all analyses are conducted on the 3D(Lo)Match dataset with YOHO (Wang et al. 2022) as the pairwise registration method.

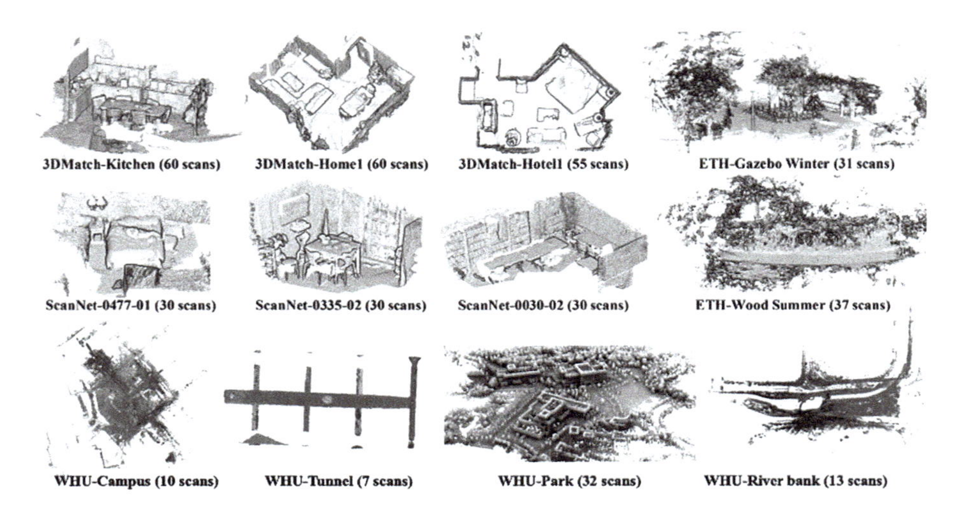

FIGURE 3.2 Qualitative results of multi-view registration on the 3DMatch, ScanNet, ETH, and WHU-TLS datasets.

3.3.1 Sparse Graph Construction

We compare the performance of EIGSE3, RotAvg~, and LITS using the fully-connected pose graph ("Full"), outliers pruned by pairwise registration results ("Pruned") and the proposed sparse graph ("Ours") in Table 3.4. It can be seen that our sparse graph construction boosts the performance of baseline methods by a larger margin than "Pruned" graphs by requiring much fewer pairwise registrations.

3.3.2 Ablation Studies on History Reweighting

We conduct ablation studies on our designs in the proposed IRLS algorithm. The results are shown in Table 3.5. We consider the following three designs.

- *Weight initialization* (*WI*). We initialize the weight to be the product of both the inlier correspondence number r_{ij} and the predicted overlap score s_{ij}. Alternatively, we may just initialize the weight with r_{ij} or s_{ij} only. Results show that the proposed initialization is better.
- *History reweighting* (*HR*). In our reweighting function, the recomputed weight is determined by rotation residuals of all previous iterations. Alternatively, we may just compute the weight from the rotation residual of current iteration. History reweighting stabilizes the iterative refinement and makes IRLS more robust to outliers.
- *Designing $g(m)$ to be increasing with m* (*INC*). In our design, we set $g(m)$ to be increasing with m so that the residuals at early iterations will have smaller impacts on results. Alternatively, we may set $g(m) = 1/M$ so that all residuals contribute equally to the weights. However, rotations estimated in the early stage are not very stable so that reducing their impacts will improve the results.

TABLE 3.4
Performances of Applying Different Multi-view Registration Methods on Different Input Pose Graphs

Pose Graph	Method	#Pair	3D-RR (%)	3DL-RR (%)
Full	EIGSE3	11905	23.2	6.6
Pruned	EIGSE3	11905	40.1	26.5
Ours	EIGSE3	**2798**	**60.4**	**44.6**
Full	RotAvg	11905	61.8	44.1
Pruned	RotAvg	11905	77.2	60.3
Ours	RotAvg	**2798**	**81.7**	**63.9**
Full	LITS	11905	77.0	59.0
Pruned	LITS	11905	80.8	65.2
Ours	LITS	**2798**	**84.6**	**68.6**

TABLE 3.5
Ablation Studies on the Proposed IRLS Scheme

	Initialization		Reweighting		
	w/o s_{ij}	w/o r_{ij}	w/o HR	w/o INC	Full
3D-RR(%)	95.5 (−0.7)	76.9 (−19.3)	83.1 (−13.1)	94.1 (−2.1)	96.2
3DL-RR(%)	79.9 (−1.7)	63.4 (−18.2)	68.9 (−12.7)	79.8 (−1.8)	81.6

3.3.3 Runtime Analysis

In Table 3.6, we provide the runtime for the graph construction and the IRLS-based transformation synchronization averaged on the 8 scenes of the 3DMatch dataset. We evaluate the runtimes on a computer with Intel(R) Core(TM) i7-10700 CPU@ 2.90GHz with GeForce GTX 2080Ti and 64 GB RAM. Our sparse pose graph construction is nearly 67s faster than baselines for conducting much fewer pairwise registrations. In total, our method is 61s ~160s faster than baselines for registering a scene in 3DMatch.

3.4 CONCLUSION

In this section, we propose a novel multi-view point cloud registration method. The key of the proposed method is a learning-based sparse pose graph construction which can estimate an overlap ratio between two scans, enabling us to select high-overlap scan pairs to construct a sparse but reliable graph. Then, we propose a novel history reweighting function in IRLS scheme, which improves robustness to outliers and has better convergence to correct poses. The proposed method demonstrates the state-of-the-art performances on both indoor and outdoor datasets with much fewer pairwise registrations.

TABLE 3.6
Detailed Time Consumption for Registering a Scene on 3DMatch

Method	Graph Cons(s)	Trans Sync(s)	Total(s)
RotAvg + Full	86.3	49.4	135.7
LITS + Full	86.3	0.7	87.0
HARA + Full	87.3	8.5	95.8
RotAvg + Pruned	164.1	22.6	186.7
LITS + Pruned	164.1	**0.7**	164.8
HARA + Pruned	164.8	7.5	172.4
Ours	**20.0**	6.9	**26.8**

Note: "Graph Cons" means the time for constructing the input pose graph. "Trans Sync" means the time for IRLS-based transformation synchronization

REFERENCES

Ao, Sheng, Qingyong Hu, Bo Yang, Andrew Markham, and Yulan Guo. 2021. "Spinnet: Learning a General Surface Descriptor for 3d Point Cloud Registration." In *Proceedings of the IEEE/CVF Conference on Computer Vision and Pattern Recognition*, 11753–62.

Arandjelovic, Relja, Petr Gronat, Akihiko Torii, Tomas Pajdla, and Josef Sivic. 2016. "NetVLAD: CNN Architecture for Weakly Supervised Place Recognition." In *Proceedings of the IEEE Conference on Computer Vision and Pattern Recognition*, 5297–307.

Arie-Nachimson, Mica, Shahar Z Kovalsky, Ira Kemelmacher-Shlizerman, Amit Singer, and Ronen Basri. 2012. "Global Motion Estimation from Point Matches." In *2012 Second International Conference on 3D Imaging, Modeling, Processing, Visualization & Transmission*, 81–8. IEEE.

Arrigoni, Federica, Beatrice Rossi, and Andrea Fusiello. 2016. "Spectral Synchronization of Multiple Views in SE (3)." *SIAM Journal on Imaging Sciences* 9 (4): 1963–90.

Chatterjee, Avishek, and Venu Madhav Govindu. 2017. "Robust Relative Rotation Averaging." *IEEE Transactions on Pattern Analysis and Machine Intelligence* 40 (4): 958–72.

Choy, Christopher, Jaesik Park, and Vladlen Koltun. 2019. "Fully Convolutional Geometric Features." In *Proceedings of the IEEE/CVF International Conference on Computer Vision*, 8958–66.

Dong, Zhen, Bisheng Yang, Yuan Liu, Fuxun Liang, Bijun Li, and Yufu Zang. 2017. "A Novel Binary Shape Context for 3D Local Surface Description." *ISPRS Journal of Photogrammetry and Remote Sensing* 130: 431–52.

Dong, Zhen, Fuxun Liang, Bisheng Yang, Yusheng Xu, Yufu Zang, Jianping Li, Yuan Wang et al. 2020. "Registration of large-scale terrestrial laser scanner point clouds: A review and benchmark." *ISPRS Journal of Photogrammetry and Remote Sensing* 163: 327–342.

Gojcic, Zan, Caifa Zhou, Jan D Wegner, Leonidas J Guibas, and Tolga Birdal. 2020. "Learning Multiview 3d Point Cloud Registration." In *Proceedings of the IEEE/CVF Conference on Computer Vision and Pattern Recognition*, 1759–69.

Huang, Shengyu, Zan Gojcic, Mikhail Usvyatsov, Andreas Wieser, and Konrad Schindler. 2021. "Predator: Registration of 3d Point Clouds with Low Overlap." In *Proceedings of the IEEE/CVF Conference on Computer Vision and Pattern Recognition*, 4267–76.

Huang, Xiangru, Zhenxiao Liang, Chandrajit Bajaj, and Qixing Huang. 2017. "Translation Synchronization via Truncated Least Squares." *Advances in Neural Information Processing Systems* 30: 1459–1468.

Kingma, Diederik P, and Jimmy Ba. 2014. "Adam: A Method for Stochastic Optimization." arXiv Preprint arXiv:1412.6980.

Lee, Seong Hun, and Javier Civera. 2022. "HARA: A Hierarchical Approach for Robust Rotation Averaging." In *Proceedings of the IEEE/CVF Conference on Computer Vision and Pattern Recognition*, 15777–86.

Pomerleau, François, Ming Liu, Francis Colas, and Roland Siegwart. 2012. "Challenging Data Sets for Point Cloud Registration Algorithms." *The International Journal of Robotics Research* 31 (14): 1705–11.

Qin, Zheng, Hao Yu, Changjian Wang, Yulan Guo, Yuxing Peng, Slobodan Ilic, Dewen Hu, and Kai Xu. 2023. "Geotransformer: Fast and Robust Point Cloud Registration with Geometric Transformer." *IEEE Transactions on Pattern Analysis and Machine Intelligence* 45 (7): 9806–9821.

Wang, Haiping, Yuan Liu, Zhen Dong, and Wenping Wang. 2022. "You Only Hypothesize Once: Point Cloud Registration with Rotation-Equivariant Descriptors." In *Proceedings of the 30th ACM International Conference on Multimedia*, 1630–41.

Yew, Zi Jian, and Gim Hee Lee. 2021. "Learning Iterative Robust Transformation Synchronization." In *2021 International Conference on 3D Vision (3DV)*, 1206–15. IEEE.

Zeng, Andy, Shuran Song, Matthias Nießner, Matthew Fisher, Jianxiong Xiao, and Thomas Funkhouser. 2017. "3dmatch: Learning Local Geometric Descriptors from Rgb-d Reconstructions." In *Proceedings of the IEEE Conference on Computer Vision and Pattern Recognition*, 1802–11.

4 Cross-Platform Point Clouds Registration

4.1 METHODOLOGY

4.1.1 Overview

To localize multiple TLS scans to ALS point clouds, the viewpoint-based descriptor named skyline context is adopted. The main idea is to simulate a subset point cloud observed from each viewpoint candidate in ALS, and then find the one most similar to a TLS scan to get an initial transformation (Böhm and Haala 2005; Cheng et al. 2018; Zang et al. 2019). Specifically, after the pre-processing of ALS and TLS, the skyline context sets of dictionaries and queries are generated. Then the best match candidate of the query in the dictionary is selected via dictionary searching and descriptor matching, and the localization is estimated and refined. The workflow is illustrated in Figure 4.1.

4.1.2 Skyline Context Generation

Given the ALS data and multiple TLS scans, the method first transfers the ALS data into a dictionary with words of skyline context and each TLS scan into a query word of skyline context. Hence, the localization of multiple scans can be solved by dictionary searching.

As the skyline context is generated by a viewpoint and its surroundings, the pre-process is to transfer the original data into a set of viewpoints and their corresponding

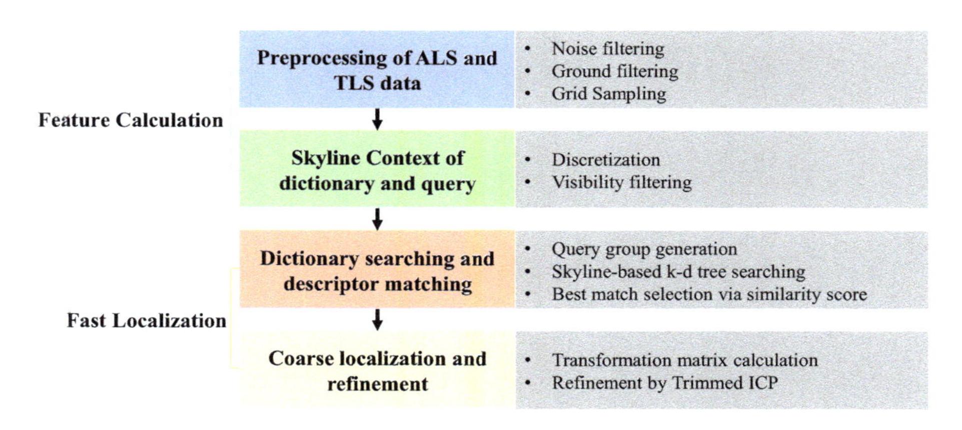

FIGURE 4.1 Workflow of skyline context for registration of multiple TLS scans to ALS. (Liang et al. 2020).

DOI: 10.1201/9781003486060-6

local point clouds. It mainly involves noise filtering, ground filtering, and grid sampling. The noise filtering is carried out by the Statistical Outlier Removal (SOR) algorithm (Rusu 2010) on both ALS and TLS. Then for the ALS, Yang's method (Yang et al. 2016) is adopted to classify the ALS data into ground points and non-ground points. The ground points are sampled as candidates with a resolution r_{ALS} (e.g., 1.0 m). For each candidate, non-ground points with a radius less than L_{max} are taken into the corresponding local point clouds. To simulate the scanning process more realistically, the candidate is moved up by assigning a virtual scanner height h_v (e.g., 1.5 m).

For TLS scans, the noise filtering contains an extra step of horizontal linear point removal. This is because the points on horizontal lines such as power lines may be rare in ALS, thus leading to a big difference in skyline shape between ALS and TLS (Yang et al. 2015). These linear points are filtered out by the linearity feature (Hackel, Wegner, and Schindler 2016). Similarly, only points with a radius less than L_{max} are retained for further process.

After the pre-processing, the candidates in ALS are used as viewpoints, together with their corresponding local point clouds to generate the skyline context dictionary words. At the same time, the TLS scan centers are used as viewpoints to generate the query words.

The second step is to generate the outdoor place descriptors(skyline contexts) from ALS and TLS as the words of the dictionary and query. To obtain the consistency of descriptors from both ALS and TLS, the impacts of resolutions and perspective views should be addressed. The steps mainly include two parts, discretization, and visibility filtering, as illustrated in Figure 4.2(a).

The discretization aims to reduce the impact of the resolution varieties (Dong et al. 2017). First, the point cloud is transformed to set the viewpoint as the origin, with its original axis orientations kept. Since points with a distance to the viewpoint greater than a certain threshold L_{max} have been removed, the local point cloud is divided into equally spaced azimuthal and radial bins, as shown in Figure 4.2(c). Suppose that N_a and N_r are the numbers of sectors and rings, then the central angle of a sector α_a and the radial gap is calculated by:

$$\begin{aligned} \alpha_a &= 2\pi / N_a \\ L_r &= L_{max} / N_r \end{aligned} \quad (4.1)$$

In this discretization mode, the equally spaced azimuthal angle and radial distance ensure that the farther the bin is, the larger its area is, reducing the effect of point density variation. After the point cloud partition, each bin is used as a basic unit to record the scene context information.

The visibility filtering is to filter the points invisible either from the top view of ALS or from the side view of TLS to retain the common points. Since a detailed visibility analysis is time-consuming, visibility filtering is accomplished in a rough and fast manner, as illustrated in Figure 4.2(b). Firstly, to filter the points invisible from the top view, only the highest point in each bin is kept. Secondly, to filter the points invisible from the scanner viewpoint, the elevation angle of each bin is calculated, and then the visibility analysis is performed on bins in the same sector. In the order of increasing radius, bins with a lower elevation angle than the previous one will be

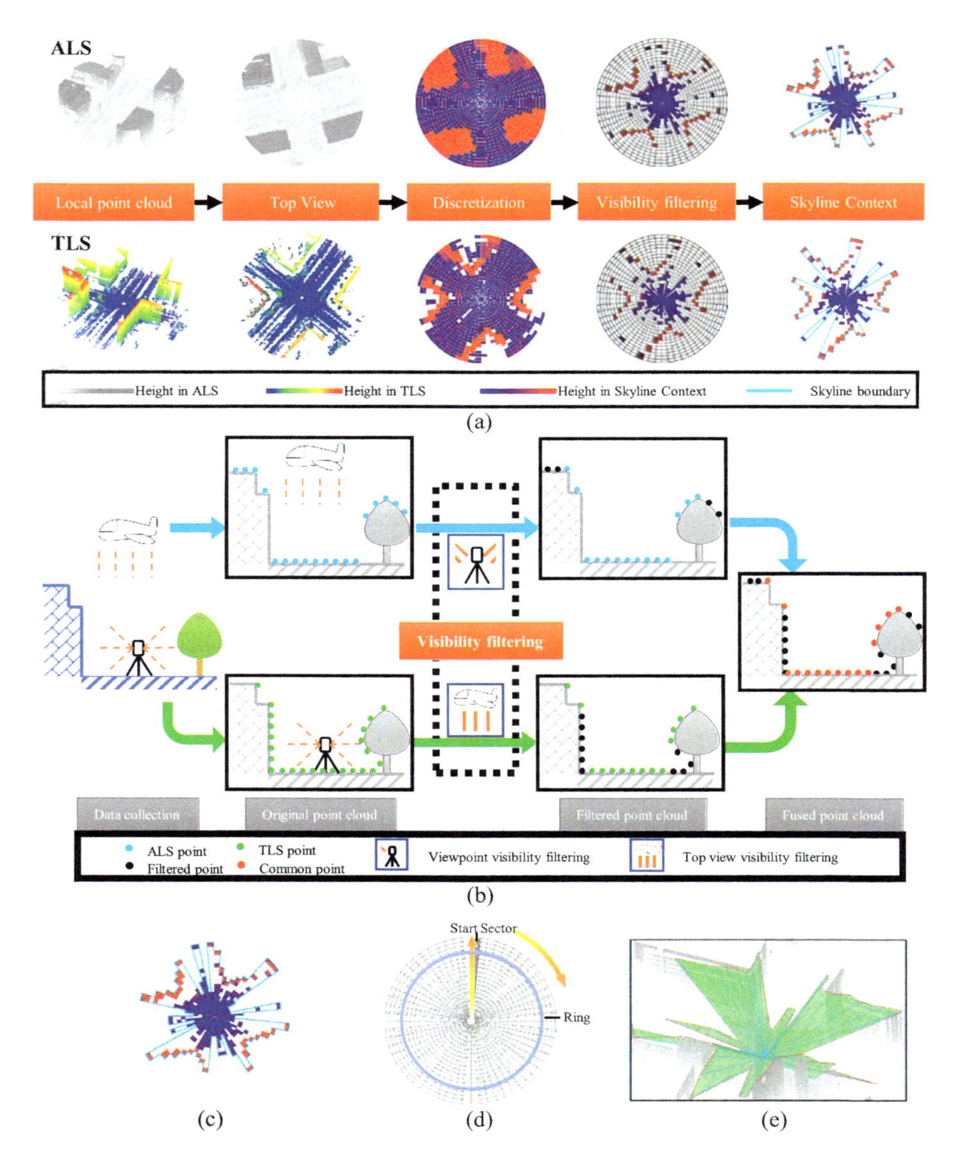

FIGURE 4.2 Generation of skyline context. (a) Examples of ALS and TLS skyline context; (b) visibility filtering schematic graph; (c) top view of Skyline Context; (d) discretization grids and recording order; (e) 3D view of Skyline Context. (Liang et al. 2020).

filtered out. Herein the field of view of the TLS scanner is used to set the valid range of elevation angle.

After discretization and visibility filtering, only points that are visible from both the top view and viewpoint are contained (Figure 4.2(c)), resulting the consistency in both ALS and TLS. The skyline context is then constructed as a series of height values in each bin with a specific order, as illustrated in Figure 4.2(d). In each sector, the height values of bins are recorded according to the radius from small to large; the sector pointing to the y-axis is set as the default start sector, followed by a clockwise adjacent sector. One $N_a * N_r$-dimension descriptor skyline context (SC) is recorded as:

$$SC = \bigcup_{j=0}^{N_a} \left\{ \bigcup_{i=0}^{N_r} h_{ij} \right\} \tag{4.2}$$

where i and j are radial and azimuthal bin numbers, and h_{ij} is the height value of the bin. The boundary of visible area connects the largest elevation angle in each sector and depicts a descriptive 3D skyline of the viewpoint, as shown in Figure 4.2(e), which is defined as the Compact Skyline Context (CSC) and written as:

$$CSC = \bigcup_{j=0}^{N_a} \left(i^*, h_{i^*j} \right), i^* = \underset{i}{argmax}\left(h_{ij} \right), h_{ij} \in SC \tag{4.3}$$

The generation of skyline context is shown in Figure 4.2. Suppose that there are N candidates in ALS, then the ALS dictionary D and the TLS query Q are written as:

$$\begin{aligned} D &= \bigcup_{n=0}^{N} SC_n^{ALS} \\ Q &= \left\{ SC^{TLS} \right\} \end{aligned} \tag{4.4}$$

4.1.3 Skyline Context-based Rapid Localization

The localization contains two parts, coarse localization and refinement. The coarse localization is to find out the most similar candidate of query from the dictionary and estimate the rotation by orientation alignment, where searching efficiency is the bottleneck.

The searching efficiency is critical as exhaustive similarity estimation in the million-level dictionary is extremely time-consuming. To effectively reduce the scope of searching, a k-d tree to index CSCs is applied. Naturally, similar SCs should have similar CSCs, and vice versa. While compared to SC, the CSC records only two values in each sector, e.g., the max of height value, and its corresponding bin distance, so that its dimension $N_a * 2$ is obviously smaller than SC's dimension $N_a * N_r$. Since a candidate word in the dictionary can be easily excluded by the obvious difference of CSC without an accurate similarity estimation, the searching efficiency can be significantly improved. The search of similar $CSCs$ is fulfilled by k-d tree.

On the other hand, the orientation alignment is necessary for rotation estimation, as the y-axis of ALS is always pointing to the north, whereas the direction of the

y-axis of TLS depends on the specific placement of the scanner at the capture time. This horizontal rotation can be approximated by the difference of the start sector between TLS and ALS. One possible solution is to set a unified start sector according to some statistical features instead of the default y-axis, for instance, setting the start sector to be the one with the largest height value or the one with the farthest valid bin. However, the statistical value-based manners are ambiguous in some special cases (e.g., there are several bins with the same height value). Therefore, a group searching approach is adopted, which transfers one query TLS skyline context SC^{TLS} into a group of query words with each sector as the start sector respectively. The most similar pairs in the group searching results are taken as the best match, and the sector difference of TLS *SC* reveals an approximate rotation angle. In this way, the computation cost inevitably increases, but it ensures that the correct match will be found in searching. Suppose Δs is the sector difference, i.e., the number of sectors from the y-axis to the shifted start sector, then the skyline context with shifted start sector and the query group QG is:

$$SC_{\Delta s}^{TLS} = \bigcup_{N_a}^{j'=0} \left\{ \bigcup_{N_r}^{i=0} h_{ij} \right\}, \; j = \left(j' + \Delta s\right) \% N_a, \; h_{ij} \in SC^{TLS} \tag{4.5}$$

$$QG = \bigcup_{\Delta s=0}^{N_{azi}} SC_{\Delta s}^{TLS} \tag{4.6}$$

In consideration of both the efficiency and the orientation problems, the skyline-based k-d tree indexing and group searching are combined for rapid localization. First, the query group of TLS *SC* is generated with the corresponding group of *CSC*s. Then the k-d tree of the dictionary is built via a skyline-based k-d tree indexing. The top K nearest neighbors (NN) of the query group are found after group searching in the k-d tree. Afterwards, the similarity scores of *SC* between the top K words and the *QG* words are calculated and sorted. The coarse localization is estimated by the best match of *SC* in the dictionary.

The similarity score function takes two *SC*s as inputs and generates a value between 0 and 1 to measure the similarity. Different from phase correlation (Avidar, Malah, and Barzohar 2017) or cosine distance (Kim and Kim 2018), we suggest a weighted height difference for the measurement, which assigns weights to each bin according to its distance to the viewpoint. Intuitively, the similarity of distant bins tends to be more important than the near ones, as a distant bin has a larger area and contains more visibility information. The similarity score *sim* of skyline context SC^{TLS} to SC^{ALS} is calculated by:

$$sim\left(SC^{TLS}, SC^{ALS}\right) = \frac{\sum_{j=0}^{N_a} \sum_{i=0}^{N_r} w_i \cdot I_{ij}^{TLS} \cdot \exp\left(-\left(h_{ij}^{TLS} - h_{ij}^{ALS}\right)^2 / 2\right)}{\sum_{j=0}^{N_a} \sum_{i=0}^{N_r} w_i \cdot I_{ij}^{TLS}} \tag{4.7}$$

where $w_i = \sqrt{i}$ is the weight of the bin, and I_{ij}^{A} is the indicator function of the bin:

$$I_{ij}^{TLS} = \begin{cases} 0, \text{ if the bin of } SC^{TLS} \text{ is empty} \\ 1, \text{ if not} \end{cases} \tag{4.8}$$

which indicates that the order of SC^{TLS} and SC^{ALS} matters in similarity estimation.

The result of coarse localization is a rotation matrix $\mathbf{R}_{coarse}$ and a translation vector $\mathbf{t}_{coarse}$. The rotation is determined by the sector difference of the best match Δs^*, and the translation is determined by the coordinate of the best match ALS point $p_n{}^*(x^*, y^*, z^*)$, where n^* is the point index. The process of query and results of coarse localization are represented by:

$$\left(\Delta s^*, n^*\right) = \underset{(\Delta s, n)}{\arg\max}\ sim\left(SC_{\Delta s}^{TLS}, SC_{n}^{ALS}\right) \tag{4.9}$$

$$\mathbf{R}_{coarse} = \begin{pmatrix} \cos\theta_z & -\sin\theta_z & 0 \\ \sin\theta_z & \cos\theta_z & 0 \\ 0 & 0 & 1 \end{pmatrix}, \theta_z = \Delta s^* \times \alpha_{azi} \tag{4.10}$$

$$\mathbf{t}_{coarse} = \left[x^*, y^*, z^*\right]^T \tag{4.11}$$

The coarse localization is further refined for a precise alignment. Trimmed-ICP (T-ICP) is chosen to address this problem. To improve the robustness, the Least Trimmed Squares (LTS) is applied in T-ICP. LTS sorts the square errors and minimizes a certain number of smaller values, so it is effective to detect the outliers from all square errors. With the outliers rejected in motion estimation, the algorithm becomes more robust (Chetverikov et al. 2002). Compared with the classic ICP, only one new parameter is required, the overlap rate of two point sets δ, which is set to 0.5 as the default value.

4.2 EXPERIMENTS

4.2.1 Datasets

Five datasets including various scenes were selected, i.e., Residence, Street, Campus, Suburban, and Highway datasets. The five datasets are considered to cover common situations in practice. All the ALS point clouds have a range of more than 1 km^2, with a point density from a minimum of 0.6 pts/m^2 (Campus) to a maximum of 97 pts/m^2 (Residence) (Dong et al., 2020). Part of the ALS data and results are shown in Figure 4.3 and Figure 4.4. All the TLS point clouds are collected in one or two years after ALS collected, containing a total of 112 TLS scans. A detailed description of the datasets is listed in Table 4.1.

The parameter settings of the method are listed in Table 4.2. The values of the parameters are invariant in processing all datasets.

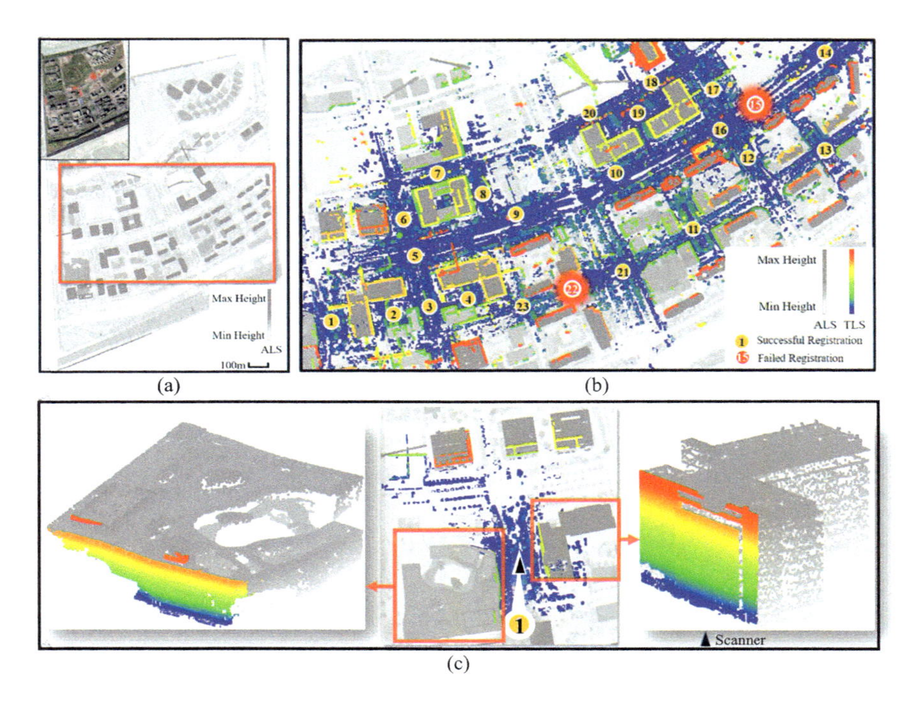

FIGURE 4.3 Street dataset and localization results. The dataset covers a range of urban areas, while the TLS data is collected along a street with a variety of architectural types (for business, office, residence), some of which have irregular shapes. (a) ALS data and corresponding satellite image of Street dataset; (b) localization result of Street dataset; (c) details at Street-1. (Liang et al. 2020).

The three parameters (N_a, N_r, L_{max}) depict the number and size of bins in generating skyline context. The parameters (N_a and N_r) greatly determine the discretization mode and impact the descriptiveness and efficiency descriptor significantly. Generally, a larger dimension tends to increase the descriptiveness of descriptors but reduces efficiency and vice versa. In this paper, the two parameters are set as (60, 20) after a thorough test with different values. The parameter L_{max} determines the radius of the neighborhood space of viewpoint and is set to 100m empirically, considering the measuring distance and the common scene size.

The parameter r_{ALS} is set as 1.0 m in the resampling of ALS data, which controls the number of candidate points and indicates a rough accuracy level of the coarse localization. The parameter h_v is an approximation of instrument height, used to unify the elevation reference of both virtual observation in ALS and actual TLS, and is set as 1.5 m.

The parameter K means to select the K nearest neighbors of a descriptor in the k-d tree, in order to quickly narrow the scope of the match. The $\boldsymbol{\delta}$ is the overlap rate of two point clouds in T-ICP. They are set as 100 and 50% respectively.

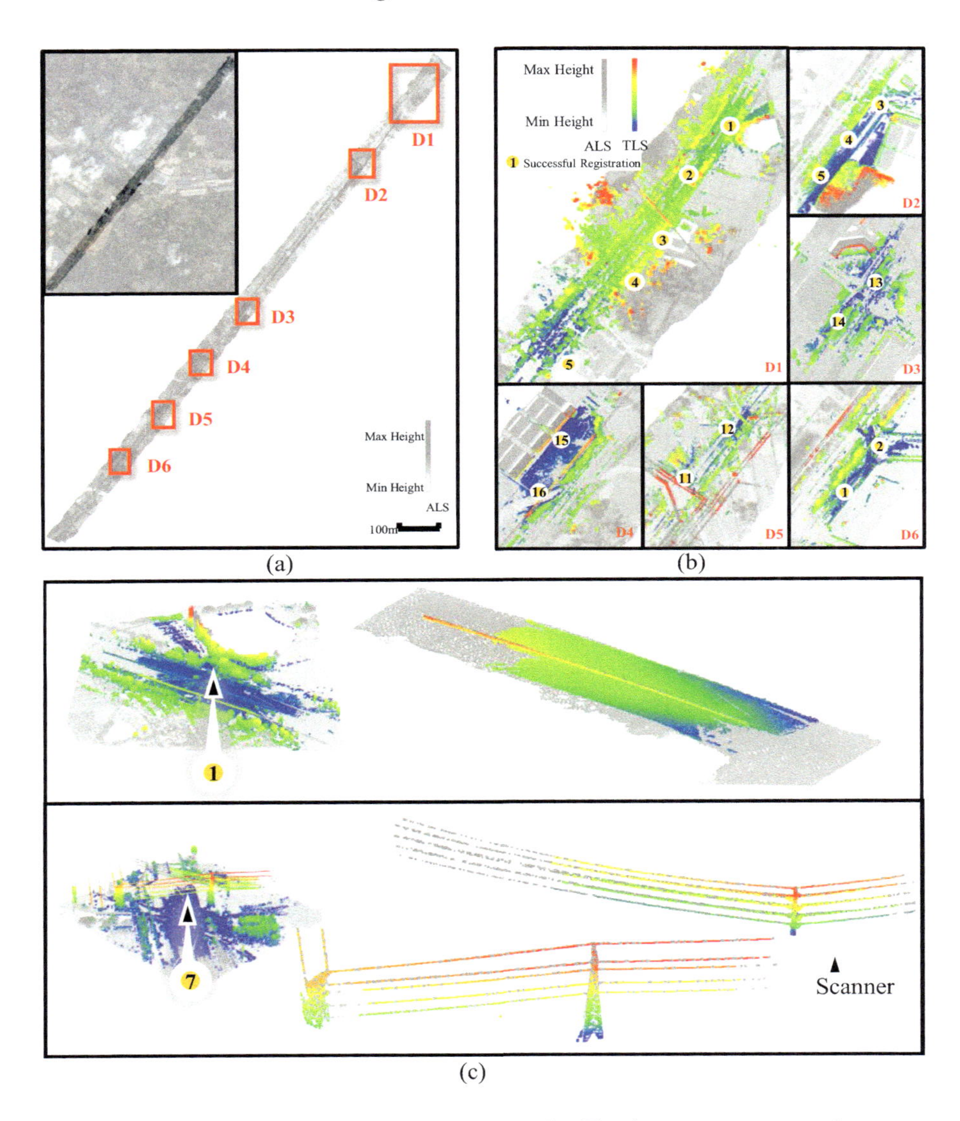

FIGURE 4.4 Highway dataset and localization results. The dataset covers a very large area with a few low-rise buildings and large amounts of vegetation and farmlands, and TLS point clouds are collected from six districts (D1-D6). (a) ALS data and corresponding satellite image of Highway dataset; (b) localization result of Highway dataset; (c) details at Highway-1 and Highway-7. (Liang et al. 2020).

TABLE 4.1
The Description of Datasets

	ALS					TLS			
	Scanner	Size (km)	#Pts (million)	Density (pts/m²)	Time (year)	Scanner	#Scans	#Pts(million)	Time (year)
Residence	Riegl VUX-1	1.3*1.3	84	97.2	2018	Z+F 5010C	30	28.8	2019
Street	Riegl VUX-1	1.2*1.6	72	49.3	2018	Riegl VZ-400	23	14.4	2019
Campus	Leica ALS70-HP	1.1*1.1	0.7	0.6	2017	Riegl VZ-400	12	12.9	2015
Suburban	Riegl LMS-Q160	1.6*0.8	3.6	3.5	2013	Riegl VZ-400	32	5.0	2014
Highway	Optech Eclipse	13.5*0.5	112	16.6	2018	Riegl VZ-400	15	23.7	2019

TABLE 4.2
Parameter Settings of the Method

Procedure	Parameter	Description	Value
Skyline Context	N_a	The number of azimuthal partition	60
	Nr	The number of radial partition	20
	L_{max}	The max range of descriptor	100 m
Generation of dictionary	r_{ALS}	The sampling resolution of ALS	1.0 m
	h_v	The virtual scanner height	1.5 m
Dictionary search	K	The number of neighbours for k-d tree searching	100
Fine registration	δ	The overlap rate for T-ICP	50%

4.2.2 Localization Quality Evaluation

With manually set ground truth, the localization results were evaluated in terms of rotation errors, translation errors, and successful localization rates (SLR). Given a pair of TLS and ALS data, the estimated transformation $\mathbf{T}_i$ and corresponding ground truth $\mathbf{T}_o$ are:

$$\mathbf{T}_i = \begin{pmatrix} \mathbf{R}_i & \mathbf{t}_i \\ 0 & 1 \end{pmatrix}$$
$$\mathbf{T}_o = \begin{pmatrix} \mathbf{R}_o & \mathbf{t}_o \\ 0 & 1 \end{pmatrix} \quad (4.12)$$

where **R** is the rotation matrix and **t** is the translation vector respectively.

Then the rotation error e_θ, translation error e_t are defined as:

$$e_\theta = \arccos\left(\frac{tr\left(\mathbf{R}_i \cdot \mathbf{R}_o^{-1}\right) - 1}{2}\right)$$
$$e_t = \| \mathbf{t}_i - \mathbf{t}_o \| \quad (4.13)$$

where $tr()$ denotes the trace of a matrix, and $\| \cdot \|$ is the L-2 norm.

A successful localization (SL) is achived when satisfying:

$$SL = \begin{cases} 1, & (e_\theta < \sigma_\theta) \wedge (e_t < \sigma_t) \\ 0, & else \end{cases} \quad (4.14)$$

where σ_θ and σ_t are pre-defined thresholds of rotation and translation error. Suppose the number of SL is N_{SL}, and the number of TLS scans is $N_{\#Scans}$, then the successful localization rate SLR is:

$$SLR = \frac{N_{SL}}{N_{\#Scans}} \quad (4.15)$$

Since this method mainly focuses on coarse localization, the *SL* and *SLR* are used for evaluation of coarse localization results with thresholds (σ_θ, σ_t) set as (5°, 3.0 m). This criterion is taken as an acceptable initial value for all refinements.

Localization results are shown in Figures 4.3 and 4.4, including a complete result of each dataset from the top view, where the TLS data is rendered by colored ribbon, and the locations of all TLS scans are marked out by corresponding numbers.

In Figures 4.3(c) and 4.4(c), a series of representative cases of each dataset is illustrated in 2D and 3D views to show more details. In Figure 4.3(c), Street-1 shows a shopping mall with irregular shapes and an office building with regular shapes, both localized well. In Figure 4.4(c), two scans Highway-1 and Highway-7 are chosen. In Highway-1, there are a lot of natural features and almost no manufactured objects. Whereas in Highway-7, the power towers and power lines are quite close to the scan center and become the main objects in the scene. As seen from the figure, the power line points from ALS and TLS fit basically overlap with each other, indicating a good localization result. In conclusion, despite many challenges, experimental results show that the method performs well on challenging environments.

The quantitative evaluation of coarse localization is listed in Table 4.3. The average rotation errors range from 0.00° to 4.47°, and translation errors from 0.10 m to 2.68 m, respectively. Since the coarse localization of all datasets is processed on the resampled ALS with a resolution of 1.0 m, their accuracy of rotation and translation of all datasets should be relatively close, which is true for all datasets except for the Campus dataset. The error of campus dataset is significantly larger due to its relatively low point density (<1 pts/m^2).

The quantitative evaluation of localization after refinement is listed in Table 4.4. The average rotation error ranges from 0.02°of street dataset to 0.89°of campus dataset. Relatively, the rotation errors of campus, suburban, and highway datasets are significantly greater than residence and street datasets. This is mainly due to that in Residence and Street dataset, there are more buildings near the viewpoint providing more angle constraints, but in the other three datasets, the point clouds are collected in more open spaces, leading to rotation errors. On the other hand, the average translation error ranges from 0.10 m in the Street dataset to 1.23 m in the

TABLE 4.3
Quantitative Evaluation of Coarse Localization

	Rotation Errors (deg)				Translation Errors (m)					
	Min	Max	Ave	RMSE	Min	Max	Ave	RMSE	#SL	SLR (%)
Residence	0.45	3.92	1.78	0.97	0.19	1.43	0.65	0.31	30/30	100%
Street	1.07	4.47	2.15	0.82	0.25	2.57	0.81	0.54	21/23	94%
Campus	0.00	4.46	1.97	1.47	0.55	2.68	1.74	0.58	11/12	92%
Suburban	0.00	3.65	1.36	0.93	0.10	2.28	0.81	0.45	31/32	97%
Highway	0.00	3.38	1.35	1.06	0.11	2.08	0.97	0.58	15/15	100%

TABLE 4.4
Quantitative Evaluation of Localization after Refinement

	Rotation Errors (deg)				Translation Errors (m)				
	Min	Max	Ave	RMSE	Min	Max	Ave	RMSE	Mean Dist (m)
Residence	0.00	0.64	0.13	0.18	0.06	0.29	0.13	0.05	0.10
Street	0.00	0.24	0.02	0.07	0.01	0.22	0.10	0.07	0.14
Campus	0.41	2.60	0.89	0.61	0.23	1.79	1.23	0.40	1.29
Suburban	0.20	1.69	0.68	0.38	0.02	1.02	0.44	0.28	0.53
Highway	0.00	1.80	0.21	0.51	0.06	0.25	0.13	0.05	0.25

Campus dataset. This error is basically affected by the point density of ALS data. For convenience, the mean distance of an ALS dataset r_{als} is listed in Table 4.4 and calculated by

$$r_{als} = \frac{1}{\sqrt{d_{als}}} \tag{4.16}$$

where d_{als} is the point density. It can be found that the average value of the translation error is less than the mean distance of the corresponding ALS data, indicating that the T-ICP achieves a good result.

4.2.3 Sensitive Analysis of Parameters

In the five datasets, a total of three failed scans (Street-15, Street-22, Suburban-11) are labeled with red numbers. These failed cases reveal some limitations of the method. For Street-15 and Street-22 in Figure 4.5(b), the trees are so close to the scanners that the generated skyline is mainly the outline of the trees, which is quite different in ALS and leads to a false localization. For Suburban-11 in Figure 4.5(c) and (d), some new trees were planted next to the location of the viewpoint after the acquisition of ALS, thus significantly changing the skyline context of TLS. This case shows that the method is not suitable for dealing with scans with significant changes during data capture.

To test the impacts of parameters (N_a and N_r) on the localization of TLS scans to ALS, the N_a is set as (40, 60, 90, 120), corresponding to the horizontal angle of (9°, 6°, 4°, 3°), and the N_r is set as (10, 15, 20, 30), corresponding to radiance distance of (10m, 6.6m, 5m, 3.3m), producing a total of 16 parameter setting cases. Four datasets including Residence, Street, Campus, and Suburban were used for the test. Among all parameter setting cases, the set of (60, 20) has the best performance of only three failures. For the 12 cases of N_r from 15 to 30 and N_a from 40 to 120, the #FL changes very little, with a maximum of eight. This reveals that the method is not sensitive to parameter settings. Whereas the #FL increases obviously in the four cases where N_r equals ten, and the failures mainly come from Residence and Suburban datasets. This is understandable because by using a larger radial gap L_{max} (10 m), the difference

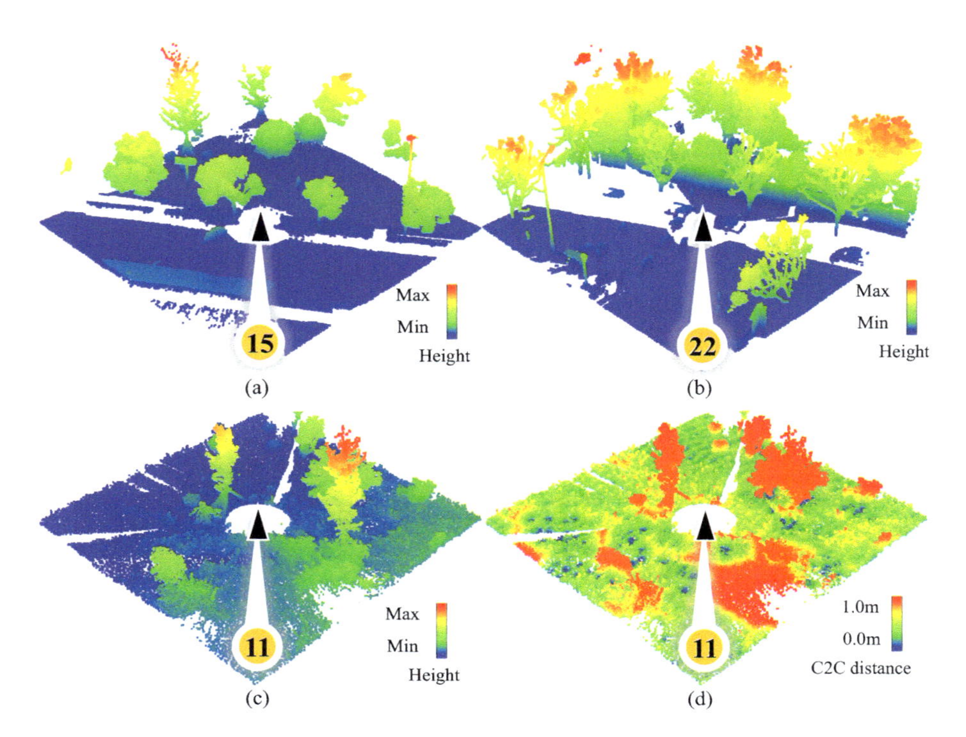

FIGURE 4.5 Failure cases. (a) Street-15, colored by height; (b) Street-22, colored by height; (c) Suburban-11, colored by height; (d) Suburban-11, colored by cloud to cloud (C2C) distance with corresponding ALS. (Liang et al. 2020).

between adjacent bins is blurred, and the discrimination of the skyline context descriptor is reduced. To be specific, in the residence dataset, the distance between adjacent buildings in District-2 is about ten meters, which is so close to the radial gap that the skyline context is unable to record a bin with ground information between the buildings. And in the Suburban dataset, this happens to the scan which is located at a narrow space between buildings and trees. Overall, the more bins the descriptor has, the stronger of descriptiveness is, but the slower of processing is. Conversely, the fewer bins the descriptor has, the higher the efficiency is, but the weaker of descriptiveness is. As a result, the parameter setting of (60, 20) is chosen as a generalized case and used in the experiments.

4.3 CONCLUSION

This chapter introduces a skyline context method for rapid localization of TLS to ALS data, producing fusion data that contains complete 3D surface information. The novel skyline context descriptor is constructed with common features from both ALS and TLS. Then a fast dictionary search by the k-d tree is implemented for coarse registration, and T-ICP is used for refinement. Experiments show that the skyline context method performs well in challenging environments such as residential areas, street canyons, campuses, suburban and highways. The method is simple and easy to implement, insensitive to parameter settings, and shows adaptability in dealing with different scenarios. Nevertheless, without the use of semantic information, the method may fail in locations where the scene has significant changes, or where the scene is too flat to generate an identifiable skyline.

REFERENCES

Avidar, David, David Malah, and Meir Barzohar. 2017. "Local-to-Global Point Cloud Registration Using a Dictionary of Viewpoint Descriptors." In *2017 IEEE International Conference on Computer Vision (ICCV)*, 891–99. Venice: IEEE. doi:10.1109/ICCV.2017.102

Böhm, Jan, and Norbert Haala. 2005. "Efficient Integration of Aerial And Terrestrial Laser Data For Virtual City Modeling Using Lasermaps."

Cheng, Liang, Song Chen, Xiaoqiang Liu, Hao Xu, Yang Wu, Manchun Li, and Yanming Chen. 2018. "Registration of Laser Scanning Point Clouds: A Review." *Sensors* 18 (5): 1641. doi:10.3390/s18051641

Chetverikov, D., D. Svirko, D. Stepanov, and P. Krsek. 2002. "The Trimmed Iterative Closest Point Algorithm." In *Object Recognition Supported by User Interaction for Service Robots*, 3: 545–48. Quebec City, Que., Canada: IEEE Comput. Soc. doi:10.1109/ICPR.2002.1047997

Dong, Zhen, Bisheng Yang, Yuan Liu, Fuxun Liang, Bijun Li, and Yufu Zang. 2017. "A Novel Binary Shape Context for 3D Local Surface Description." *ISPRS Journal of Photogrammetry and Remote Sensing* 130 (August): 431–52. doi:10.1016/j.isprsjprs.2017.06.012

Dong, Zhen, Fuxun Liang, Bisheng Yang, Yusheng Xu, Yufu Zang, Jianping Li, Yuan Wang et al. "Registration of large-scale terrestrial laser scanner point clouds: A review and benchmark." *ISPRS Journal of Photogrammetry and Remote Sensing* 163 (2020): 327–342.

Hackel, Timo, Jan D. Wegner, and Konrad Schindler. 2016. "Contour Detection in Unstructured 3D Point Clouds." In *2016 IEEE Conference on Computer Vision and Pattern Recognition (CVPR)*, 1610–18. Las Vegas, NV, USA: IEEE. doi:10.1109/CVPR.2016.178

Kim, Giseop, and Ayoung Kim. 2018. "Scan Context: Egocentric Spatial Descriptor for Place Recognition Within 3D Point Cloud Map." In *2018 IEEE/RSJ International Conference on Intelligent Robots and Systems (IROS)*, 4802–9. Madrid: IEEE. doi:10.1109/IROS.2018.8593953

Liang, Fuxun, Bisheng Yang, Zhen Dong, Ronggang Huang, Yufu Zang, and Yue Pan. 2020. "A Novel Skyline Context Descriptor for Rapid Localization of Terrestrial Laser Scans to Airborne Laser Scanning Point Clouds." *ISPRS Journal of Photogrammetry and Remote Sensing* 165 (July): 120–32. doi:10.1016/j.isprsjprs.2020.04.018

Rusu, Radu Bogdan. 2010. "Semantic 3D Object Maps for Everyday Manipulation in Human Living Environments." *KI - Künstliche Intelligenz* 24 (4): 345–48. doi:10.1007/s13218-010-0059-6

Yang, Bisheng, Ronggang Huang, Zhen Dong, Yufu Zang, and Jianping Li. 2016. "Two-Step Adaptive Extraction Method for Ground Points and Breaklines from Lidar Point Clouds." *ISPRS Journal of Photogrammetry and Remote Sensing* 119 (September): 373–89. doi:10.1016/j.isprsjprs.2016.07.002

Yang, Bisheng, Yufu Zang, Zhen Dong, and Ronggang Huang. 2015. "An Automated Method to Register Airborne and Terrestrial Laser Scanning Point Clouds." *ISPRS Journal of Photogrammetry and Remote Sensing* 109 (November): 62–76. doi:10.1016/j.isprsjprs.2015.08.006

Zang, Yufu, Bisheng Yang, Jianping Li, and Haiyan Guan. 2019. "An Accurate TLS and UAV Image Point Clouds Registration Method for Deformation Detection of Chaotic Hillside Areas." *Remote Sensing* 11 (6): 647. doi:10.3390/rs11060647

5 Point Clouds and Panoramic Images Registration

5.1 METHODOLOGY

To obtain the correspondence between the panoramic image and the point cloud, an automatic registration method by pole matching is applied. This method mainly includes pole instance segmentation in the panorama, pole object extraction from the point cloud, pole matching, and registration model calculation, as described in Figure 5.1.

5.1.1 Pole Instance Segmentation on the Panorama

For pole semantic segmentation, a neural network (Andrew et al., 2020) is adopted, which combines hierarchical, multi-scale attention mechanisms and context features to achieve high-precision image semantic segmentation while needing more computation. Considering the limitation of GPU computation, we split the panoramic image into several parts, and then merge all the results after the semantic segmentation of each part, to achieve a full semantic segmentation. Finally, all pole objects are extracted according to the semantic label.

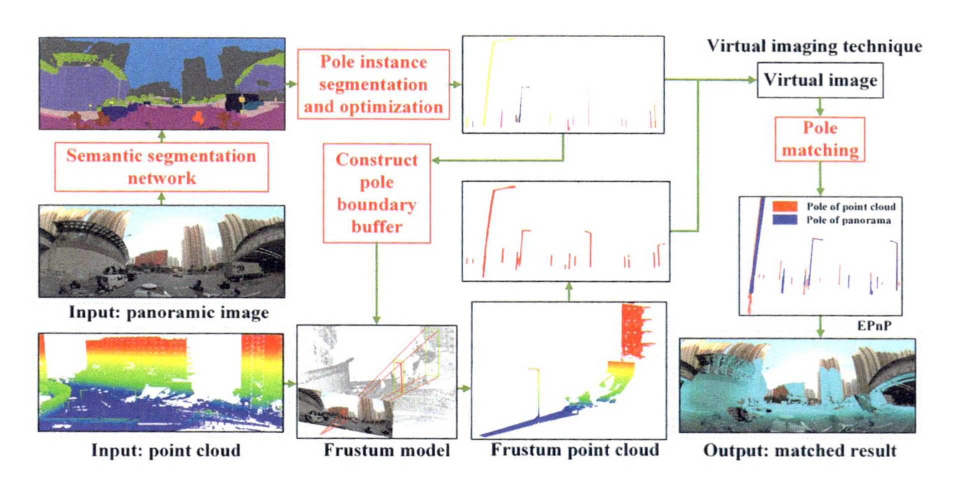

FIGURE 5.1 Architecture of the method.

DOI: 10.1201/9781003486060-7

To acquire the pole instances, the segmentation result is transformed into a binary image (pixels belonging to the pole are 1, others are 0). With the occlusion of other objects such as billboards, signal lights, cars, and pedestrians, some poles are discontinuous in the vertical direction. It is difficult to deal with the occlusion problem with the region-growing algorithm (Tremeau and Borel, 1997), so we improve the search strategy. The solution lies in limiting the neighborhood search space of the seed pixel (x, y) to the vertical direction, which is denoted as Eq. (5.1):

$$H = \left\{ \bigcup_{i=y}^{\beta} (x,i) \mid P(x,\beta) = 1 \, and \sum_{j=\beta+1}^{imgHeight} P(x,j) = 0 \right\} \tag{5.1}$$

where $\bigcup_{i=y}^{\beta} (x,i)$ indicates the pixels obtained from the growth of the seed pixel along the vertical direction, $P(\cdot)$ refers to the pixel value, $\beta \in [y, H_I]$ shows which row the last pole pixel along the vertical direction lies in. For seed pixel (x, y), search along the vertical direction using the region growing algorithm, and get the last pole pixel (x, β), and then set the value of all the pixels between (x, y) and (x, β) along the vertical direction as 1. Finally, the discontinuous poles are reconnected in the vertical direction. Thus, the segmentation results are refined to complete the pole binary mask, which is instantiated by connected domain analysis to extract all the pole instances on the panoramic image.

5.1.2 Pole Object Extraction from Point Cloud

Since the point cloud in the urban scene is cluttered, it is hard to accurately extract pole objects in the whole point cloud scene, and the process is time-consuming. To improve the accuracy and efficiency, the frustum point cloud (Qi et al., 2018; Su et al., 2020) is constructed by establishing the correspondence between the point cloud and panoramic image through the initial extrinsic orientation parameter. Considering the inaccuracy of parameters, a shape-adaptive buffer region of pole instance on panorama is built to ensure the completeness of the pole point cloud. Hence, only pole objects in the frustum need to be extracted, which reduces the complexity of the scene, and improves the accuracy and efficiency.

5.1.2.1 Frustum Point Cloud

According to the initial pose, the point cloud is transformed into the spherical coordinate system through Eq. (5.2), and then the spherical coordinates are transformed into the pixel coordinates of the panoramic image by Eq. (5.3) and (5.4). Finally, the mapping between the point cloud and the panoramic image is constructed, and therefore all the points whose projection results fall within the 2D bounding box of the pole instance form a frustum point cloud. Combining the perspective model with the pole instance in a panoramic image, the 3D pole object extraction can be transformed from the whole scene to the frustum point cloud, which eliminates the noise and significantly improves the accuracy and efficiency.

$$\begin{pmatrix} x_p \\ y_p \\ z_p \\ 1 \end{pmatrix} = \begin{pmatrix} R^{initial} & T^{initial} \\ 0 & 1 \end{pmatrix} \begin{pmatrix} x_w \\ y_w \\ z_w \\ 1 \end{pmatrix} \tag{5.2}$$

$$r = \sqrt{x_p^2 + y_p^2 + z_p^2}, \theta = \tan^{-1}\left(\frac{y_p}{x_p}\right), \varphi = \cos^{-1}\left(\frac{z_p}{r}\right) \tag{5.3}$$

$$\theta = \pi - \frac{x}{W_I} * 2\pi, \varphi = \frac{y}{H_I} * \pi \tag{5.4}$$

where (x_w, y_w, z_w) indicates the coordinate of the point cloud, (x_p, y_p, z_p) indicates the intersection of the line between (x_w, y_w, z_w) and the center of the sphere, $R^{initial}$ and $T^{initial}$ are the initial extrinsic orientation parameter, θ and φ are the angles between (x_p, y_p, z_p) and the x-axis and z-axis respectively, x and y are columns and rows in the panoramic image corresponding to the coordinates (x_p, y_p, z_p) on the sphere, H_I and W_I are the height and width of the panoramic image.

5.1.2.2 Point Cloud Classification

Affected by the deviation of the initial pose parameter, the pixel coordinate $P_{initial} = \{(x, y)|x \in [0, W_I], y \in [0, H_I]\}$ projected by the point cloud onto the panoramic image is inaccurate. As a consequence, the corresponding complete pole object of the point cloud cannot be directly acquired by using the pole instance in the panoramic image. Thus a shape-adaptive buffer region is constructed for the bounding box of the pole instance, $P_{buffer} = \{(x, y)|x \in [0, W_I + d_1], y \in [0, H_I + d_2]\}$, where d_1 and d_2 are the shape-adaptive value of the buffer size, which is determined by the distance from the camera exposure position. By expanding the boundary of the pole instance in the panoramic image, the corresponding frustum point cloud contains the complete pole object.

Although the search space is transformed from the whole point cloud scene to the frustum point cloud, there are still a large number of background points. Therefore, slicing, clustering, and connected domain analysis methods are combined to extract pole objects. Firstly, the ground points and human-made planes are eliminated (Fan et al., 2020). Secondly, the rest point clouds are sliced at the z-axis direction, and then, the Euclidean distance clustering method is used for each slice. Thirdly, the classes in the clustering results that satisfy the cylindrical parameters are retained, and the others are excluded. Finally, the candidate cylinders that are discontinuous in the vertical direction are eliminated, and the rest are part of the poles. The point cloud is divided into two classes, pole (red) and others (blue). For each pole instance in the panoramic image, a frustum point cloud is generated, which contributes to the extraction of all pole objects. After that, all the detected pole objects in every frustum are fused.

5.1.3 Pole Matching and Registration Model Calculation

5.1.3.1 Virtual Imaging

The panoramic image is essentially a spherical model. If the spherical model is directly expanded into a plane, there will be significant deformation. The geometric structure of objects in panoramic images is distorted. This chapter presents a pole-matching method based on virtual images (Chen et al., 2013; Li et al., 2018).

Assuming that the coordinate of a point on the virtual imaging plane is $(u, v, -f)^T$, the imaging point coordinate of the spherical model is (x_p, y_p, z_p), the correspondence pixel position on the expanded model is (x, y), the collinearity equation between the image points on the panoramic sphere and the virtual image points is denoted as Eq. (5.5).

$$\begin{bmatrix} u \\ \mathrm{v} \\ -f \end{bmatrix} = \lambda R_v \begin{bmatrix} x_p \\ y_p \\ z_p \end{bmatrix} = \lambda R_v \begin{bmatrix} \sin\left(\dfrac{y}{imageHeight}\pi\right)\sin\left(\pi - \dfrac{x}{imageWidth}*2\pi\right) \\ \sin\left(\dfrac{y}{imageHeight}\pi\right)\cos\left(\pi - \dfrac{x}{imageWidth}*2\pi\right) \\ \cos\left(\dfrac{y}{imageHeight}\pi\right) \end{bmatrix} \tag{5.5}$$

where x and y are the column and row numbers of the current pixel on the panoramic image, R_v is the camera rotation matrix of the virtual imaging method, and the f and λ are the focus and scale factors respectively. Combined with the pre-defined interior orientation parameter, pixel size, and the virtual camera size, through the inverse solution of formula (5), the virtual image with any focus and orientation can be acquired.

5.1.3.2 Particle Swarm Optimization (PSO)

After virtual images of 2D and 3D poles are obtained, the two virtual images are converted into corresponding binary images (the value of the pole is 1, others are 0), which are denoted as I_{pc} and I_{pano}, respectively. The loss function is Eq. 5.6.

$$L = \| I_{pc} + d - I_{pano} \|_0 \tag{5.6}$$

where $\| \cdot \|_0$ is the L_0 norm and d is the offset of the point cloud pole projection image. Obviously, after transformation, when the overlap between the binary image is the highest, L is the smallest.

Based on the initial parameters, the search space is limited to a local range. Thus, we use particle swarm optimization (Wang et al., 2018) to calculate the optimal transformation parameters. For the i_{th} particle, the candidate solution is r_i^k, the current speed is v_i^k, and at the same time, r_i^g preserves the optimal estimate of the loss function before the K generation. In addition, for all particles, r^g preserves the optimal estimate of the overall loss function, so the optimization formula of speed in each iteration is calculated as Eq. (5.7).

$$v_i^k = v_i^{k-1} + c_1 \cdot rand(0,1) \cdot \left(r_i^g - r_i^{k-1}\right) + c_2 \cdot rand(0,1) \cdot \left(r^g - r_i^{k-1}\right) \tag{5.7}$$

where c_1 and c_2 are individual and overall learning factors, which are set to 3.0 and 1.5 respectively according to a large number of experimental results, *rand*(0, 1) represents the uniform random number between (0, 1), and the updated formula of the optimal solution is shown in Eq. (5.8).

$$r_i^k = r_i^{k-1} + v_i^k \tag{5.8}$$

5.1.3.3 Transformation Calculation

After applying PSO, the optimal offset d is obtained, which enables an extensive correspondence between point cloud and virtual image. To obtain accurate transformation parameters, all corresponding points need to be evenly distributed in the whole image. The image is divided into N parts, and one pair of correspondence is selected in each part. Finally, with enough 2D-3D points obtained, EPnP (Vincent et al., 2009) is used to calculate the accurate rotation (R) and translation parameters (T) from the point cloud to the panoramic image, achieving fine registration between the point cloud and the panoramic image. According to Eq. (5.3), Eq. (5.4), and Eq. (5.9), the MLS point cloud can be accurately transformed onto the plane of the panoramic image.

$$\begin{bmatrix} x_w \\ y_w \\ z_w \end{bmatrix} = \lambda R^{-1} R_v \left(\begin{bmatrix} x_p \\ y_p \\ z_p \end{bmatrix} + T \right)$$

$$= \lambda R^{-1} R_v \left(\begin{bmatrix} \sin\left(\frac{y}{imageHeight}\pi\right) \sin\left(\pi - \frac{x}{imageWidth} * 2\pi\right) \\ \sin\left(\frac{y}{imageHeight}\pi\right) \cos\left(\pi - \frac{x}{imageWidth} * 2\pi\right) \\ \cos\left(\frac{y}{imageHeight}\pi\right) \end{bmatrix} + T \right) \tag{5.9}$$

where R_v is the camera rotation matrix of the virtual imaging method, R and T are calculated by EPnP.

5.2 EXPERIMENTS

In this section, we analyze the performance of our method using the urban datasets from Hong Kong and Shanghai, as shown in Figure 5.2. Table 5.1 lists the details of the datasets. These scenes are urban street scenes with a great number of pole objects. According to the frequency of the sensor, we select a panoramic image every five frames and the corresponding point cloud for testing. The performance was tested with NVIDIA GTX 1080Ti, Intel i7-9700H CPU, and 32GB Memory.

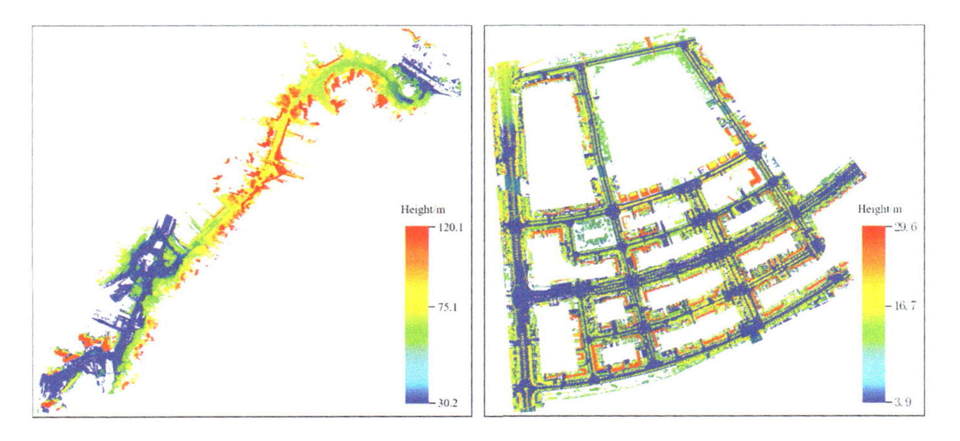

FIGURE 5.2 Point cloud datasets of urban scenes, with color indicating different heights.

TABLE 5.1
Data Description

Parameters	Hong Kong	Shanghai
Number of the panoramic image	60	100
Resolution of pixel	8000*4000	8192*4096
Number of points	115500672	175182745
Length(m)	1500	8000

5.2.1 Pole Extraction and Matching Results on Images and Point Clouds

Owing to the occlusion of other objects and the influence of light, some pole objects are separated vertically. After optimizing the separated pole object by the modified region growing algorithm, a more complete one is obtained.

Finally, the extracted pole object is instanced by using the connected domain analysis method. By using slicing, clustering, and connected domain analysis in the frustum point cloud, the extraction speed and accuracy of the 3D pole objects are extremely improved.

Figure 5.3 illustrates the matching results of Hong Kong and Shanghai urban scenes, where blue and red are the projection results of 2D and 3D poles respectively. The left column indicates the relative positions of the poles before matching and the right column is the matching results.

In Figure 5.3b, the number of 2D pole objects is larger than that of 3D pole objects. This is because some pole objects are not collected during point cloud collection. In the pole-matching process, if the 3D pole object corresponding does not exist, the 2D pole object will not be applied for pole matching. Therefore, the inconsistency between the number of 2D and 3D pole objects does not affect the final registration.

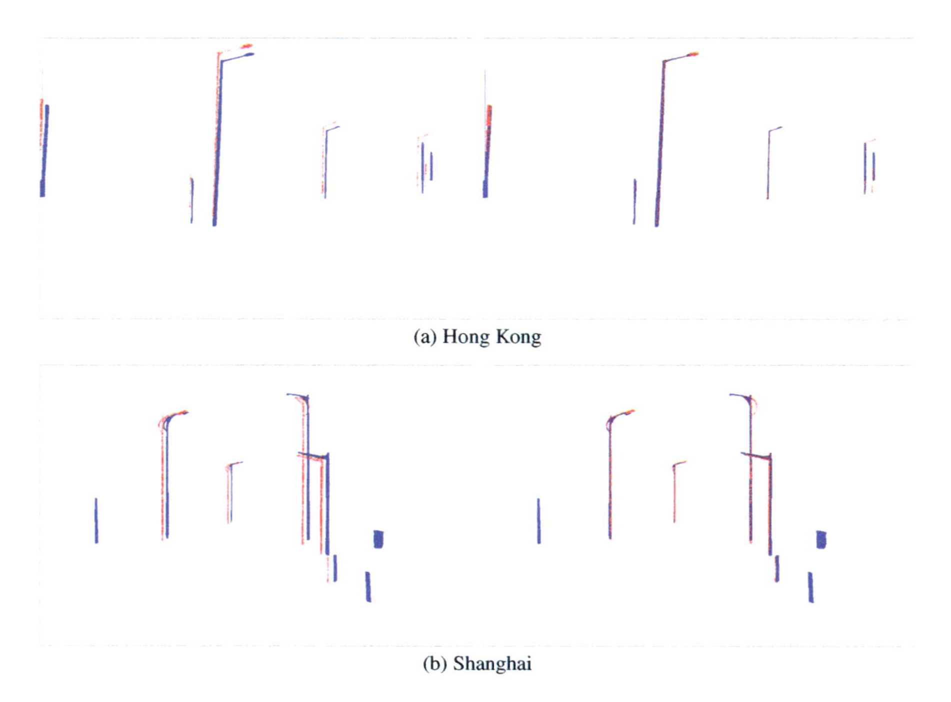

(a) Hong Kong

(b) Shanghai

FIGURE 5.3 Matching results of Hong Kong and Shanghai urban scenes. The left column indicates the relative positions of the poles before matching and the right column is the matching results.

In addition, to deal with the semantic segmentation errors in the panoramic image, two cases are considered. First, the pole is wrongly recognized as other objects. In this case, the 2D pole object is not extracted, nor will the corresponding 3D pole object, which leaves no effect on the matching result. Second, other objects are wrongly segmented into poles. In this case, a 3D pole cannot be extracted from the corresponding frustum point cloud. Therefore, the fault will not affect the registration result as well. The method applies panoramic images and point clouds as constraints to each other, thus improving the accuracy and robustness.

5.2.2 Registration Results

The final registration results are shown in Figure 5.4. Figure 5.4a and 5.4b show the registration effect of the point cloud and the panoramic image of the urban scene in Hong Kong and Shanghai, respectively. The first and second rows are the registration results using the initial pose and the calculated parameters respectively, and the two columns on the right are the details. A large registration error of the point cloud and panoramic image lies in the first row. Meanwhile, significant improvement is achieved by the method, with almost complete overlap.

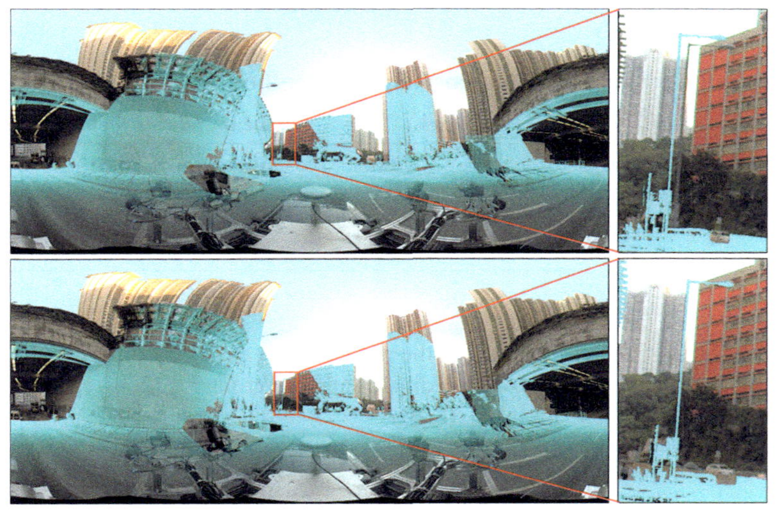

(a) Hong Kong urban scene

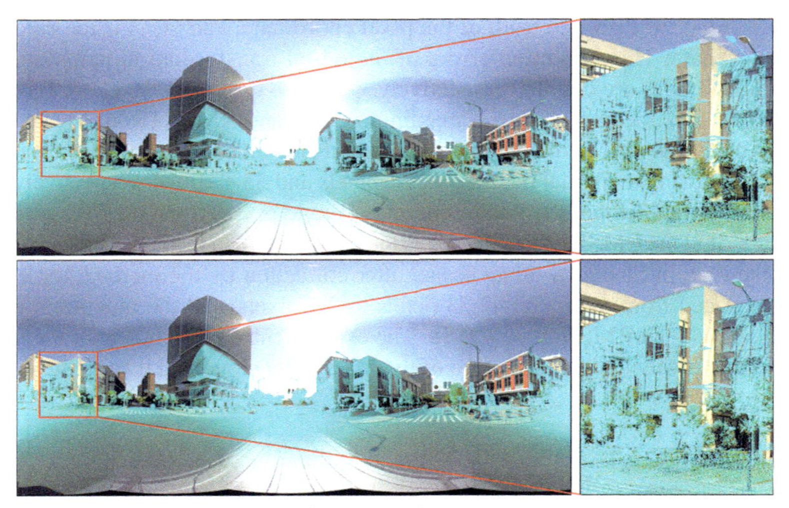

(b) Shanghai urban scene

FIGURE 5.4 Registration results. The first line shows the registration based on the initial parameter. The second line shows the result using the parameter calculated by our method.

5.3 ANALYSIS AND DISCUSSION

5.3.1 Quantitative Analysis

To quantitatively evaluate the performance, the checkpoints in the point cloud and panoramic image are manually chosen as the ground truth. To ensure the reliability of the ground truth, the intersection, corner, and other easily distinguishable points are preferred. Then we use the transformation parameters to project the point cloud

TABLE 5.2
Registration Accuracy of Initial Pose and the Method

Scene	Initial Pose (pixels)			The Introduced Method (pixels)		
	Average	Max	RMSE	Average	Max	RMSE
Hong Kong	33.47	42.31	4.77	2.35	5.05	1.03
Shanghai	43.33	52.17	5.00	2.10	4.28	0.88

to the panoramic image for collecting the pixel coordinates. In addition, the offset of the calculation between the pixel coordinates and the ground truth is regarded as the registration error, as denoted in Eq. (5.10).

$$Pixel_{\text{error}} = \sqrt{\left(x - x^{gt}\right)^2 + \left(y - y^{gt}\right)^2} \tag{5.10}$$

where (x^{gt}, y^{gt}) is the ground truth of the pixel coordinate, and (x, y) is the calculated pixel coordinate. The projection errors of the initial transformation parameters and the calculated ones are counted respectively.

Table 5.2 lists the average error, maximum error, and RMSE (Root Mean Square Error) of the initial pose and method. In the Hong Kong urban scene, the maximum and average error using the initial pose are 42.31 pixels and 33.47 pixels respectively. After correction by the method, the maximum and average pixel errors are 5.05 pixels and 2.35 pixels, and the RMSE of the method is much smaller than the initial pose, with only 1.03. Likewise, in Shanghai urban scene, after correction by the method, the average error is reduced from the initial 43.33 pixels to 2.1 pixels, and the maximum error is 4.28 pixels, with an RMSE of only 0.88. The experiment results show that the method performs effectively in the registration of the MLS point cloud and panoramic image.

5.3.2 Ablation Experiment

5.3.2.1 Pole Object Extraction from Point Cloud

To verify the accuracy of extracting poles from combined point clouds and panoramic images, a comparison is made between using only point clouds and combining point clouds and panoramic images.

To quantitatively evaluate the effect of extracting pole objects by two methods, we take the pole objects in the panoramic image as the reference. The extraction is successful if it matches the number of pole instance objects in the panoramic image. The buffer of the pole instance is used in the frustum point cloud extraction, so the actual number of pole objects in the point cloud may be larger than in the panoramic image, which does not affect the pole-matching module and so still considered successful. Whereas, if the number of extracted pole objects is relatively small, the extraction is dropped. The successful registration rate (SRR) and efficiency are used as the evaluation criteria for this experiment. The final results are shown in Table 5.3. In the two

TABLE 5.3
Comparison of SRR and Efficiency in 3D Pole Object Extraction

	Combine Point Cloud and Panoramic Image		Only Point Cloud	
Scene	**SRR**	**Efficiency (s/frame)**	**SRR**	**Efficiency (s/frame)**
Hong Kong	92%	15.73	80%	122.28
Shanghai	96%	22.87	84%	150.19

datasets, the SRR of the method combining the point cloud and the panoramic image is 92% and 96%, respectively, while the SRR of only using the point cloud is just 80% and 84%. Furthermore, the efficiency of the method combining the point cloud and the panoramic image is 15.73 s/frame and 22.87s/frame, respectively, which is seven times faster than that of using only the point cloud. Experiments show that the method combining point cloud and panoramic image achieves superior performance in extracting 3D pole objects.

5.3.2.2 Shape-adaptive Buffer Region

To validate the effectiveness of the shape-adaptive buffer region, a comparison is carried out, with a fixed-size buffer region. The fixed-size buffer region is generally set to be large enough for all cases, especially for long-range objects, because the deviation of long-range objects is more obvious. The SRR and efficiency of 3D pole object extraction are adopted as the evaluation criteria. The quantitative comparison is shown in Table 5.4.

In Table 5.4, The SRR of the two methods to extract poles are equal. It indicates that both types of buffer regions meet the requirement to include the complete 3D pole objects. On the other hand, the buffer region obtained by the shape-adaptive method is normally smaller than the fixed-size one, so the extraction efficiency of 3D pole objects is improved. As a result, the efficiency of the fixed-size buffer region method in Hong Kong and Shanghai is 19.39s/frame and 33.22s/frame respectively,

TABLE 5.4
Comparison of SRR and Efficiency between Shape-adaptive Buffer Region and Fixed-size Buffer Region

	Shape-adaptive		Fixed-size	
Scene	**SRR**	**Efficiency (s/frame)**	**SRR**	**Efficiency (s/frame)**
Hong Kong	92%	15.73	92%	19.39
Shanghai	96%	22.87	96%	33.22

lower than the efficiency of the shape-adaptive buffer region method, which is 15.73s/frame and 22.87s/frame respectively.

5.3.3 Discussion

5.3.3.1 Comparative Evaluation

Zhu et al. (2021) proposed the Relative Orientation Model (ROM) to register the MLS point cloud and panoramic image sequences and verified it with the urban scene data of Shanghai. They used ten panoramic images to test, taking the fifth image as the initial frame, manually selecting the control point, and using DLT to calculate the transformation parameters, achieving a minimum error of 4.01 pixels. For other frames, the transformation parameters were obtained indirectly by matching with adjacent panoramic images, leading to a worse registration accuracy of panoramic images far away from the initial image, with a maximum error of 8.45 pixels. Therefore, to better compare the registration results, we select the registration results of the first, sixth, and tenth panoramic images according to the distance for comparison and display. The results are shown in Figure 5.5.

For the sixth panoramic image, both methods perform well, and the average error is around four pixels. For the first panoramic image, as the distance increases, the error of ROM begins to accumulate, resulting in a larger error. Our method achieves a higher accuracy in registration results. For the last panoramic image, the error

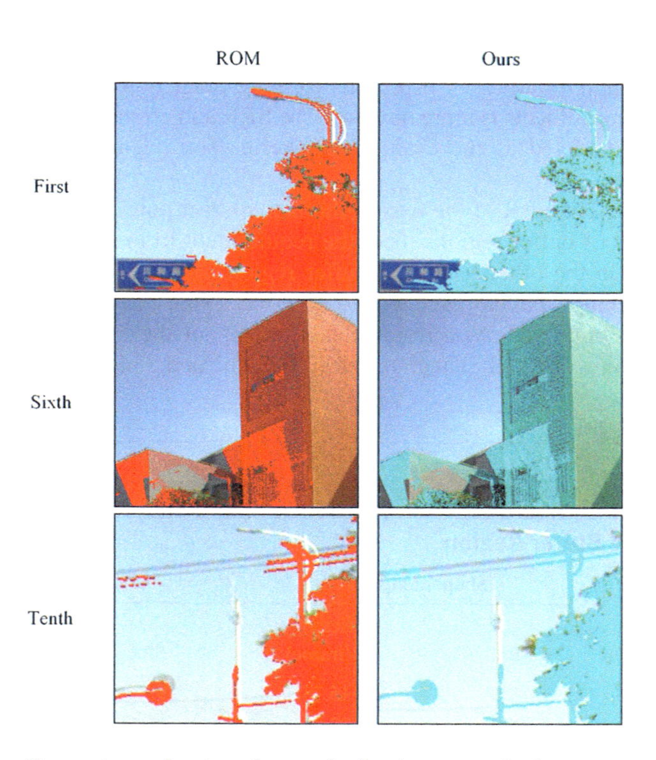

FIGURE 5.5 Comparison of registration results for the two methods.

TABLE 5.5
Registration Accuracy (pixels) Comparison of ROM and Our Method

	1	2	3	4	5	6	7	8	9	10	Mean	RMSE
ROM	7.64	7.75	4.75	3.84	4.05	4.01	6.27	4.86	6.16	8.45	5.77	1.63
Ours	4.50	3.45	4.66	2.91	3.36	3.22	2.90	2.88	5.25	3.36	**3.54**	**0.84**

accumulation of ROM reached the maximum, leading to the worst registration result with 8.45 pixels. In contrast, our method achieves superior performance with a registration accuracy of 3.36 pixels.

To quantitatively estimate the registration accuracy of the two methods, Table 5.5 lists the average pixel error and RMSE of ROM and our method. Obviously, in the test datasets, the performance of our method is superior. The maximum error of our method is 5.25 pixels, which is much smaller than the 8.45 pixels of ROM. Likewise, the average error and RMSE of our method are 3.54 pixels and 0.84 respectively, which are smaller than the 5.77 pixels and 1.63 of ROM.

5.3.3.2 Analysis of Failure Case

The method is based on the matching of pole objects for the registration of point clouds and panoramic images. Therefore, the number of poles directly affects the registration performance. Two failure cases are found in the experiment results, both because of the limited pole objects in the scene.

In addition, the EPnP is used to solve the transformation parameters, and thus, the distribution of 2D-3D corresponding points should follow the rules of non-collinear, non-coplanar, and uniform distribution. However, in the experimental results, there are scenes where the selected corresponding points are coplanar and unevenly distributed, resulting in registration failure. Two poles are extracted from the point cloud and the panoramic image respectively, and a large number of 2D-3D corresponding points can be acquired for the matching. However, in the panoramic image, only part of the lamp pole is extracted. It can be seen that all 2D-3D points selected on the two-pole objects are coplanar, leading to registration failure. Therefore, when there are only two poles in the scene, non-coplanar correspondences are necessary for registration. Generally, when there are three or more poles in the scene, the adopted correspondences are naturally not coplanar.

In summary, for scenes where few or no poles exist, or only collinear poles are available, the method may fail. To address these cases, traditional methods like the ROM can still be considered as a reference.

5.4 CONCLUSION

The registration of mobile laser scanning point clouds and panoramic images is important for many applications, such as point cloud classification and instance segmentation, robot navigation, augmented reality, high-definition map construction, and 3D reconstruction. The unforeseen sensor movement after calibration, the interference of the environment during driving, and temporal misalignment during the data collection, may lead to errors in the matching between the point cloud and the panoramic image. In this chapter, a novel pole-matching method is introduced to accurately and robustly register the point cloud and panoramic image. Firstly, a semantic segmentation network is adopted to extract the 2D pole object in the panoramic image. Secondly, the optimized 2D pole object with a shape-adaptive buffer region is applied to generate the frustum point cloud, and then the 3D pole object is extracted from those points. Thirdly, the 2D and 3D pole objects are transformed into virtual images. A novel pole instance matching algorithm is applied that maximizes the overlapping areas via solving the conjugate primitive finding problem by PSO to obtain sufficient 2D-3D correspondences, and then the transformation parameters are calculated by the EPnP. The method performs effectively on two challenging urban scenes, obtaining the average registration error of 2.01 pixels (with RMSE 0.88) and 2.35 pixels (with RMSE 1.03), respectively. However, the method is not applicable in scenarios where there are no valid pole objects due to occlusion, sensor data collection, etc. In the future, linear features from trees or buildings can be considered as registration primitives in the model to increase the robustness of the method.

REFERENCES

Andrew Tao, Karan Sapra, and Bryan Catanzaro, 2020. "Hierarchical multi-scale attention for semantic segmentation." arXiv preprint arXiv:2005.10821.

Chen Chi, Bisheng Yang, and Chaoyi Deng, 2013. "Registration of the Panoramic Image Sequence onto Mobile Laser Scanning Point Cloud Using Linear Features." *The International Symposium on Mobile Mapping Technology (MMT2013).*

Fan Wen, Yang Bisheng, Liang Fuxun, and Dong Zhen. 2020. "Using Mobile Laser Scanning Point Clouds to Extract Urban Roadside Trees for Ecological Benefits Estimation." *The International Archives of the Photogrammetry, Remote Sensing and Spatial Information Sciences*, 43: 211–16. doi: 10.5194/isprs-archives-xliii-b2-2020-211-2020

Li Jianping, Bisheng Yang, Chi Chen, Ronggang Huang, Zhen Dong, and Wen Xiao. 2018. "Automatic Registration of Panoramic Image Sequence and Mobile Laser Scanning Data Using Semantic Features." *ISPRS Journal of Photogrammetry and Remote Sensing* 136: 41–57. doi: https://doi.org/10.1016/j.isprsjprs.2017.12.005

Pang Su, Morris Daniel, and Radha Hayder. 2020. "CLOCs: Camera-LiDAR Object Candidates Fusion for 3D Object Detection." *IEEE/RSJ International Conference on Intelligent Robots and Systems (IROS).* 10386–10393. doi: 10.1109/IROS45743.2020.9341791

Qi Charles R., Wei Liu, Chenxia Wu, Hao Su, and Leonidas J. Guibas. 2018. "Frustum PointNets for 3D Object Detection from RGB-D Data." *IEEE Conference on Computer Vision and Pattern Recognition*, 918–27. doi: 10.1109/cvpr.2018.00102

Tremeau Alain, and Nathalie Borel. 1997. "A Region Growing and Merging Algorithm to Color Segmentation." *Pattern Recognition* 30: 1191–1203. doi: 10.1016/s0031-3203(96)00147-1

Vincent Lepetit, Moreno-Noguer Francesc, and Fua Pascal, 2009. "EPnP: An Accurate O(n) Solution to the PnP Problem." *International Journal of Computer Vision* 81: 155–66. doi:10.1007/s11263-008-0152-6

Wang Dongshu, Dapei Tan, and Lei Liu. 2018. "Particle Swarm Optimization Algorithm: An Overview." *Soft Computing* 22: 387–408. doi: 10.1007/s00500-016-2474-6

Zhu Ningning, Yang Bisheng, Dong Zhen, Chen Chi, Huang Xia, and Xiao Wen. 2021. "Automatic Registration of Mobile Mapping System Lidar Points and Panoramic-Image Sequences by Relative Orientation Model." *Photogrammetric Engineering & Remote Sensing*, 87(12): 913–22. doi: 10.1016/j.isprsjprs.2017.12.005

Part III

Detection and Segmentation of Point Cloud

The rapid advancement of digitalization and automation in transportation, urban planning, and management has necessitated the development of algorithms capable of accurately interpreting and interacting with complex three-dimensional environments. 3D point clouds collected by LiDAR (Light Detection and Ranging) offer a detailed representation of the physical world. Effectively processing point clouds and gaining exact information is a challenge in the intelligent understanding of the 3D environment. This part will synthesize the key contributions and insights of three distinct yet complementary aspects in point cloud analysis: 3D object detection, semantic segmentation, and instance segmentation.

This part first introduces a cutting-edge model for 3D object detection from point clouds, a fundamental task in autonomous driving. Traditional approaches, divided into anchor-based or anchor-free models, have struggled with the inaccuracies introduced by occlusions and the limited detection range of LiDAR systems. These challenges are encountered when attempting to infer 3D bounding boxes for objects, as the physical center of the object often lacks corresponding measured 3D points. We introduce a novel corner-guided anchor-free single-stage 3D object detection model (CG-SSD) to overcome these limitations. This model employs a 3D sparse convolution backbone network, coupled with a corner-guided auxiliary module (CGAM), to enhance the feature representation of objects, especially those that are small or partially occluded. The CG-SSD model has demonstrated state-of-the-art performance on benchmarks like ONCE, showcasing significant improvements in accuracy and offering a promising solution for the autonomous driving industry.

Building upon the foundation of accurate object detection, we continue to delve into point cloud semantic segmentation, which is essential for comprehensive 3D scene understanding. We recognize the unique challenges in large-scale outdoor point cloud segmentation, such as sample imbalance, huge data volume, object scale variation, and complex structures. To tackle these issues, we propose an end-to-end network with an effective spatial downsampling strategy, a point-based feature abstraction module, and a tailored loss function to mitigate category imbalance. This network, evaluated on the Toronto-3D and Shanghai MLS datasets, achieves impressive mIoU scores, highlighting its superiority in segmenting urban scenes and its potential for enhancing urban planning and management.

Based on semantic segmentation, we tackle the point cloud instance segmentation task. Outdoor point clouds lack appearance and texture information, posing challenges in distinguishing object instances for detailed scene understanding. Therefore, we integrate street-view imagery with point clouds to segment road infrastructure in street scenes. This pipeline involves generating 2D proposals from street object detection models, predicting instance masks and bounding boxes for each 3D frustum, and fusing frustums from multiple perspectives using multi-object tracking. The method has been validated on datasets from Shanghai and Wuhan, achieving high mean recall and precision rates and demonstrating robustness across different urban environments.

In summary, this part collectively represents approaches in the field of point cloud processing, providing innovative solutions to distinct challenges, from object detection to semantic and instance segmentation. It contributes to a deeper understanding of how algorithms can reliably interpret and interact with the intricate geometries of our built environments. The insights and methodologies presented push the boundaries of current technology and open up new avenues for future research and practical applications in autonomous driving, digital twins, and urban informatics.

6 3D Object Detection

6.1 METHOD

This chapter introduces the CG-SSD model. CG-SSD consists of voxelization, 3D and 2D feature learning, CGAM, and detection head. First, CG-SSD assigns the point clouds to the regular 3D grids and extracts voxel features using 3D sparse convolution. After getting the 3D voxel features, CG-SSD projects the features to the Bird's Eye View(BEV) and uses the 2D backbone to extract deeper features. Unlike other networks, CG-SSD adds CGAM between the 2D backbone and detection head module. Then, the corner features extracted by the auxiliary network and the features obtained by the 2D backbone are fused, and finally, the classification and regression tasks of the object are performed. Figure 6.1 shows the architecture of the CG-SSD.

6.1.1 Backbone

Voxelization. To deal with sparse and irregular point cloud data, there are methods based on the point, voxel, and projection to learn point cloud features. Compared with other feature learning methods, the voxel-based 3D sparse convolution method learns features with non-empty voxels and can reduce time and resources for computation (Dai, Diller, and Niebner 2020; Yan, Mao, and Li 2018). In addition, 3D sparse

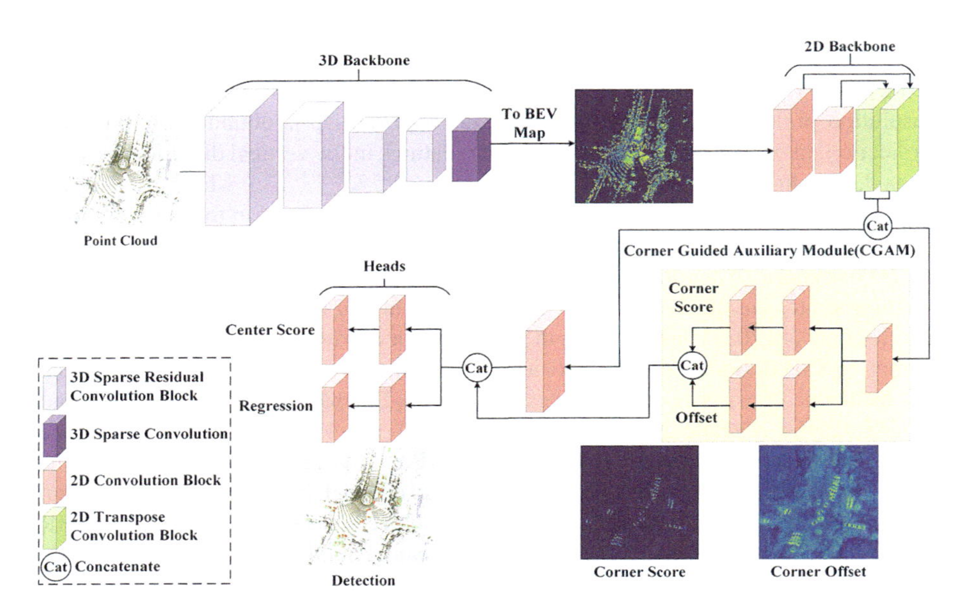

FIGURE 6.1 The structure of the CG-SSD model (Ma et al. 2022).

DOI: 10.1201/9781003486060-9

convolution can get richer voxel features with different kernel sizes and strides. In CG-SSD, the voxel-based feature learning method is selected. Given a point cloud dataset $P = \{p_0, p_1, \ldots, p_n\} \subseteq \mathcal{R}^m$, point cloud range $[x_{min}, y_{min}, z_{min}, x_{max}, y_{max}, z_{max}]$ and voxel size $[v_x, v_y, v_z]$, the voxel index of a point p_i with coordinate (x, y, z) and feature $f \in \mathcal{R}^{m-3}$ is located according to Equation 6.1.

$$Index_i = \left[int\left(\frac{x_i - x_{min}}{v_x}\right), int\left(\frac{y_i - y_{min}}{v_y}\right), int\left(\frac{z_i - z_{min}}{v_z}\right) \right] \subseteq \mathcal{R}^3. \tag{6.1}$$

Due to the sparsity of the point clouds, the number of points within each voxel is different. Thus, CG-SSD sets a fixed value T. If the number of points belonging to the voxel is greater than T, then T points are randomly sampled from all the points of the voxel. If the number of points is less than T, CG-SSD pads to T with zeros. After such an operation, CG-SSD gets N non-empty voxels, forming a three-dimensional tensor: $N \times T \times m$. CG-SSD will use 3D sparse convolution to perform feature extraction on this Tensor.

3D backbone. To efficiently learn the features of 3D voxels, CG-SSD adopts 3D Sparse Convolution (Yan, Mao, and Li 2018) as the 3D backbone. The structure of this module is a backbone composed of residual layers and sub-manifold sparse convolutional layers. After the multi-layer convolutions, each voxel can learn spatial features at different scales, and finally, obtain voxels with richer features. CG-SSD uses the same parameters of the 3D backbone as the CenterPoints (Yin, Zhou, and Krähenbühl 2021) and SECOND (Yan, Mao, and Li 2018). After three convolutions with stride = (2, 2, 2), kernel = (3, 3, 3), padding = (1, 1, 1), the size of each voxel becomes $[8 \cdot v_x, 8 \cdot v_y, 8 \cdot v_z]$. To simplify object classification and regression tasks, 3D voxel features are then projected to BEV according to the spatial indices of voxels. In order to reduce the amount of calculation, the 3D sparse convolution with stride = (2, 1, 1), kernel = (3, 1, 1), padding = (0, 0, 0) is finally used to reduce the resolution in the vertical direction. Since the feature map is obtained by 3D voxel projection, in order to avoid feature loss, the features in the vertical direction are connected. The final projected BEV feature map is $I \subseteq \mathcal{R}^{H \times W \times (D \times F)}$, where H, W is the size of the projected feature map and D is the number of voxels in the vertical direction and F is the dimension of voxel features.

2D backbone. After obtaining the BEV feature map, CG-SSD also adopts the same learning strategy as other state-of-art models (Hu et al. 2021; Shi et al. 2020; Yin, Zhou, and Krähenbühl 2021), that is, using the 2D convolution layers to learn deeper features. Here CG-SSD designs an SSD-like(Liu et al. 2016) network as the 2D backbone. The 2D backbone consists of multiple sets of Conv2D-BatchNorm2D-ReLU (CBR) and ConvTranspose2D-BatchNorm2D-ReLU (CTBR) blocks. Firstly, following the SECOND (Yan, Mao, and Li 2018) and CenterPoints (Yan, Mao, and Li 2018), the CBR blocks with stride as 1, kernel size as 3*3, and channel number as 128 are adopted, focusing on learning the characteristics of small objects. Secondly, by using the CBR block with stride as 2, kernel size as 3*3, and channel numbers as 256, the resolution of the feature map is reduced and the receptive field of the convolution kernel is increased, to learn the features of the large objects. Thirdly, CG-SSD uses 2D transpose convolution layers with stride as 1 and 2, and channel number as 256 respectively, so that the size of the output 2D features is identical, and the final output of the 2D backbone is the combination of the output of CTBR blocks. More experiments are provided in the experiment section to demonstrate that the rationality of the 2D backbone structure. Through the 2D backbone, the network can learn different levels of features, which can be better used for object classification and detection of different sizes.

6.1.2 Corner-guided Auxiliary Module (CGAM)

Most of the recent models directly perform object classification and regression after 3D and 2D feature learning. However, due to the influence of factors such as occlusion between objects and the distance from the target to the sensor, it is difficult for the sensor to obtain complete targets' surface point clouds. It is unfavorable for the network to learn the object center location from incomplete point clouds. Therefore, CG-SSD supervises the network to learn more accurate center points by predicting the corners of the 3D bounding boxes on the BEV map, thereby improving the detection accuracy of the network.

6.1.2.1 Corner Selection Strategy

As shown in Figure 6.2, on the BEV map, if the sensor can observe the object, it must be able to observe a corner (Visible Corner, VC) of its bounding box. Since many object point clouds are distributed around the corner, the network can easily learn the position of the corner. However, the other three kinds of corners (Partly-Visible Corner L (PVCL), Partly-Visible Corner W (PVCW), and Invisible Corner (IVC)), which are hard to observe by the sensor, should be given more attention. By learning the positions of these corners, the model is potentially guided to aggregate the target features towards the center point.

The ground truth 3D bounding boxes of the datasets (KITTI (Geiger et al. 2013), ONCE (Mao et al. 2021), Waymo (Sun et al. 2020)) are composed of the center point position, dimension (length, width, height), and rotation angle. To supervise the network to learn corner features, during the training process, CG-SSD needs to calculate

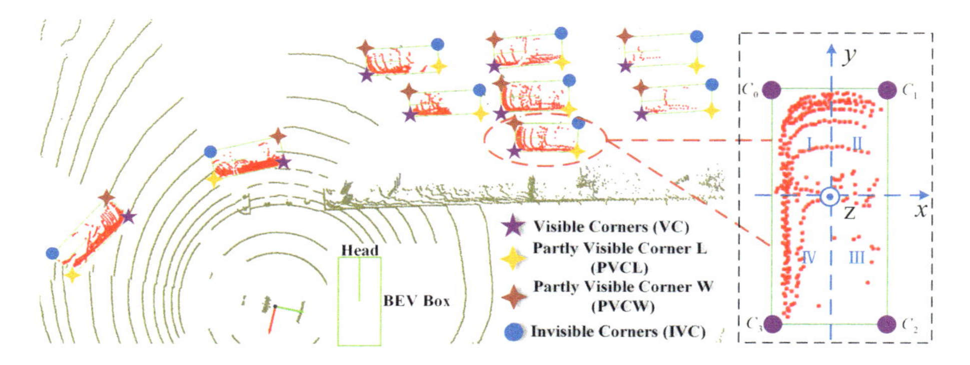

FIGURE 6.2 The visibility of object corners on the BEV map (Ma et al. 2022).

ALGORITHM 1. THE ADAPTIVE CORNER SELECTION ALGORITHM

Data: n points with local coordinate $P \subseteq R^{n*3}$, points in 3D bounding boxes index $P_{index} \subseteq R^{n*1}$, 3D bounding boxes number m, corners of 3D bounding boxes′ top surface $C \subseteq R^{m*4*3}$, index used to select auxiliary corners $index$ = [(2, 3, 1), (3, 2, 0), (0, 1, 3), (1, 0, 2)]

Result: selected corners C_f, C_l, C_w for auxiliary module

1. $C_f \leftarrow zeros(m, 2)$
2. $C_l \leftarrow zeros(m, 2)$
3. $C_w \leftarrow zeros(m, 2)$
4. **for** $i \leftarrow 0$ **to** m **do**
5. $p_i \leftarrow P_{index} = = i$;
6. $X \leftarrow P[p_i, 0]$;
7. $Y \leftarrow P[p_i, 1]$;
8. $q_0 \leftarrow sum(X < 0 \text{ and } Y > 0)$;
9. $q_1 \leftarrow sum(X > 0 \text{ and } Y > 0)$;
10. $q_2 \leftarrow sum(X > 0 \text{ and } Y < 0)$;
11. $q_3 \leftarrow sum(X < 0 \text{ and } Y < 0)$;
12. $q \leftarrow [q_0, q_1, q_2, q_3]$;
13. $sub\_q \leftarrow [(q_0 + q_1 + q_3), (q_1 + q_0 + q_2), (q_2 + q_1 + q_3), (q_3 + q_2 + q_0)]$;
14. $valid_q \leftarrow sum(q > 0)$;
15. **if** $valid_q \leq 2$ **then**
16. $max_i \leftarrow argmax(q)$;
17. **else**
18. $max_i \leftarrow argmax(sub\_q)$;
19. **end**
20. $C_f[i] \leftarrow C[i, index[maxi, 0]]$;
21. $C_l[i] \leftarrow C[i, index[maxi, 1]]$;
22. $C_w[i] \leftarrow C[i, index[maxi, 2]]$;
23. **end**

the position of each visible corner and the other three corners. However, because the orientation of the target in the LiDAR coordinate system is different, and the density of points in the target box is different, the location of each category corner cannot be determined. Therefore, CG-SSD applies an adaptive corner selection algorithm based on the number of points in the 3D bounding box. First, CG-SSD converts the LiDAR coordinate system to the local coordinate system through Equation 6.2, to obtain the points' coordinates in the 3D bounding box in local coordinates.

$$\begin{vmatrix} x_i' \\ y_i' \\ z_i' \end{vmatrix} = \begin{vmatrix} x_i - x_c \\ y_i - y_c \\ z_i - z_c \end{vmatrix} \begin{vmatrix} \cos(\theta) & -\sin(\theta) & 0 \\ \sin(\theta) & \cos(\theta) & 0 \\ 0 & 0 & 1 \end{vmatrix} \quad i = 0,1,2,\ldots,m \tag{6.2}$$

where (x_i, y_i, z_i) is the coordinate in the LiDAR coordinate system, (x_c, y_c, z_c) is the object center in the LiDAR coordinate system, θ is the object orientation angle, m is the number of points, and $\left(x_i', y_i', z_i'\right)$ is points' coordinate in local coordinate. As shown on the right of Figure 6.2, the center of the object is taken as the origin, the front direction of the object is the positive direction of the $y-$ axis, and the positive direction of the $x-$ axis is the direction in which the $y-$ axis rotates 90° clockwise. As described in Algorithm 1, according to the $x-y$ axis, CG-SSD divides the object's points into four quadrants and counts the number of points in each quadrant. The corner corresponding to the quadrant with the most points is visible (VC in Figure 6.2), and it is the easiest to detect by the network, while the remaining three corners are likely to be observed by the sensor. The corner opposite the visible corner is invisible (IVC in Figure 6.2), and the other two corners are partly visible (PVCL and PVCW in Figure 6.2).

6.1.2.2 Auxiliary Module

In the previous model (Hu et al. 2021; Yin, Zhou, and Krähenbühl 2021), after obtaining more advanced features through the 2D backbone, the head module is used for object detection. To obtain more accurate detection results, CG-SSD added a corner-guided auxiliary module (CGAM) before the head module (Figure 6.3). The network learns the edge information and rotation information of the object by detecting partially visible and invisible corners and obtains a more accurate center point position, dimension, and angle. This module is mainly composed of some CBR blocks, which are supervised by calculating the losses of classification and position offset during network training.

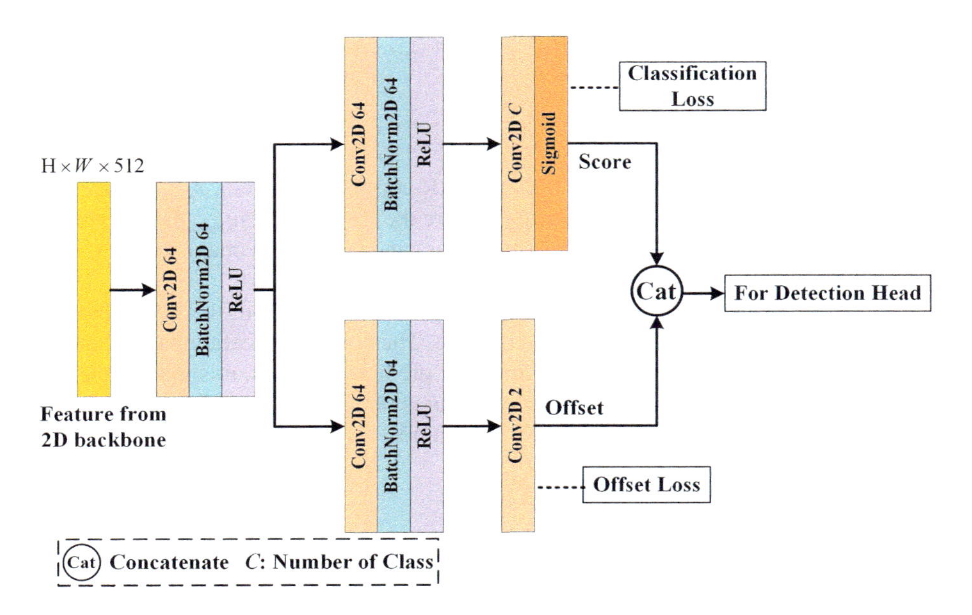

FIGURE 6.3 Corner-guided auxiliary module (Ma et al. 2022).

In the classification task of the CGAM, the size of the final output of the model is $H \times W \times C$, where H and W are the sizes of the score map, and C is the number of classes. For each object, only the pixel where the corner is located is positive, and the rest are negative. The target assignment strategy will bring about an unbalanced distribution of positive and negative samples, which affects network training. Therefore, CG-SSD adopts the same strategy as CornerNet (Law and Deng 2020), that is to reduce the penalty of pixels near the corners. The 2D Gaussian value (Equation 6.3) is set for the pixel which is located within a circle with the corner as the center and r as the radius.

$$p = e^{-\frac{x^2+y^2}{2\sigma^2}} \tag{6.3}$$

where x and y are the coordinate differences between the pixel and corner in the BEV map, $\sigma = \frac{r}{3}$, and p is the target value. In this chapter, r is set to 2.

In the regression task of the auxiliary module, the final output of the model is $H \times W \times 2$, whose third dimension represents the offset between the pixel center's coordinate and the corner's coordinate on the BEV map. Due to the voxel size setting, row and column numbers of the corners in the BEV map are decimals. Therefore, there is an offset when using the row and column numbers to express the precise position of the corner. And the offset can be calculated according to Equation 6.4–6.6. The regression branch in CGAM is used to predict the deviation value.

$$p_x, p_y = \left(\frac{x_c - x_{min}}{v_x \cdot out_{factor}}, \frac{y_c - y_{min}}{v_y \cdot out_{factor}} \right) \tag{6.4}$$

$$x_o = x_c - \left[int(p_x) \cdot v_x \cdot out_{factor} + x_{min} \right] \tag{6.5}$$

$$y_o = y_c - \left[int(p_y) \cdot v_y \cdot out_{factor} + y_{min} \right] \tag{6.6}$$

where x_c, y_c are the corner's value in the LiDAR coordinate system, x_{min}, y_{min} are the minimal value of inputted point clouds, v_x, v_y are the voxel size, out_{factor} is the ratio of the voxel size before and after the 3D backbone, and is 8 in this chapter. The x_o and y_o are the target value.

Additionally, the score map, regression map, and feature from the 2D backbone are concatenated for the detection head module. The combined feature has a channel with $f + (C + 2) \cdot n$, where C is the number of predicted classes, n is the number of corners for each object, f is the number of features obtained by the 2D backbone and set to 512, and 2 represents the corner offset value.

6.1.3 Detection Head and Loss Function

6.1.3.1 Detection Head

The detection head module is designed for classification and regression tasks with enhanced features. In the classification branch, the strategy of center assignment and Equation 6.3 are used to obtain the ground-truth value of network training. In the regression branch, the model outputs the offset of the center point, the dimension, and the orientation angle of objects, respectively. Only the pixel corresponding to the center point position is assigned the true value, and other pixels are assigned zeros. In the process of network training, CG-SSD gets such true value according to Equation 6.7.

$$x_t = x_o, y_t = y_o, z_t = z,$$
$$w_t = \log(w), l_t = \log(l), h_t = \log(h), \tag{6.7}$$
$$a1 = \sin(\theta), a2 = \cos(\theta)$$

where x_o and y_o are the offset value of the object's center, z is the center point's coordinate in the vertical direction, (w, l, h) are the dimensions (width, length, height), and θ is the orientation angle. The detection head module of the CG-SSD model is composed of multiple CBR blocks. First, one CBR block transforms the features from the 2D backbone. Second, both the classification and regression branches consist of one middle CBR block and a Conv2D layer to get the predictions of the network. The input of the classification and regression branches is the concatenation of the features learned by the CBR block and the features from the auxiliary module. Consistent with the detection head of CenterPoints (Yin, Zhou, and Krähenbühl 2021), the stride of all CBR layers is 1, the kernel is 3, and the out channel is 64 in this module.

6.1.3.2 Multi-task Loss

In the CG-SSD model, the classification loss function is a variant of Focal Loss (Law and Deng 2020), and the regression loss function is L_1 Loss. The focal loss can be calculated by Equation 6.8.

$$L_f = \frac{-1}{N_c} \sum_{c=1}^{C} \sum_{i=1}^{H} \sum_{j=1}^{W} \begin{cases} (1-p_{cij})^{\alpha} \log(p_{cij}) & if \ y_{cij} = 1 \\ (1-y_{cij})^{\beta} (p_{cij})^{\alpha} \log(1-p_{cij}) & otherwise \end{cases} \tag{6.8}$$

where N_c is the number of positive samples, C is the number of classes, H and W are the sizes of score map, y_{ij} is the ground truth value of classification, the p_{cij} is the network prediction. Following CornerNet (Law and Deng 2020), the hyperparameters α, β are set to 2, 4. The L_1 Loss is used in the regression module in the CG-SSD model, which can be calculated by Equation 6.9,

$$L_g = \frac{1}{N_r} \sum_{i=1}^{N_r} \left| G_i^t - G_i^p \right| \tag{6.9}$$

where N_r is the number of positive pixels in the regression map, G^p is the prediction of the regression branch, and G^t is the ground truth value. Especially, in the auxiliary module, $G^t \in [x_o, y_o]$ and x_o, y_o are the offset value of the corner from Equation 6.5–6.6.

And in the detection head module, $G^t \in [x_t, y_t, z_t, w_t, l_t, h_t, a_1, a_2]$ which is calculated from Equation 6.7.

The total loss L of the CG-SSD model is:

$$L = L_{center\_cls} + \gamma \cdot L_{center\_reg} + \lambda \cdot L_{aux} \tag{6.10}$$

$$L_{aux} = L_{corner\_cls} + L_{corner\_reg} \tag{6.11}$$

where L_{center_cls} is the focal loss of the detection head module, L_{center_reg} is the L1 loss of regression branch in the detection module, L_{aux} is the loss of the corner-guided auxiliary module, which is the sum of the classification and regression loss of the corner, and γ, λ are the loss weight, which are set to 0.25, 0.25 respectively.

6.2 EXPERIMENTS

6.2.1 Dataset

The CG-SSD model is tested on the ONCE Dataset (Table 6.1) (Mao et al. 2021). ONCE dataset was published for 3D object detection in the autonomous driving, scenario, consisting of one million LiDAR scenes. And the LiDAR data is acquired by a 40-beam LiDAR sensor in multiple cities in China. ONCE dataset contains six training sequences (5k scenes), four validation sequences (3k scenes), and ten testing sequences (8k scenes) for model evaluation. During the evaluation, it merges the car, bus, and truck into a super-class "vehicle". The official evaluation metric for ONCE dataset is orientation-aware mean average precision (mAP), which is obtained by averaging the AP of three categories. To ensure fair comparisons, CG-SSD sets a detection point cloud range of [− 75.2m, 75.2m] for x and y axis, and [− 5.0m, 3.0m] in the vertical direction.

To demonstrate that CGAM can be integrated as a plug-in to most networks using BEV features, CG-SSD also evaluated it on the Waymo Open Dataset (Sun et al. 2020). Waymo contains 798 training sequences, 202 validation sequences, and 150 testing sequences with vehicles, pedestrians, and cyclists. The LiDAR data is captured with a 64-beam LiDAR sensor. In this chapter, CG-SSD only predicts the object in the point cloud range of [− 75.2m, 75.2m] for x and y axis, and [− 2.0m, 4.0m] in the vertical direction. Following the open-source project OpenPCDet[1], CG-SSD only trains the model on 20% of the training set and evaluates it on the full validation set.

TABLE 6.1
The Dataset Used in the CG-SSD Model

Dataset	Publish Year	Scenes	Classes	Area
ONCE	2021	1M	car, bus, truck, pedestrian, cyclist	China
Waymo	2019	230k	vehicle, pedestrian, cyclist, sign	US

6.2.2 Network Details

Data Augmentation. Following SECOND (Yan, Mao, and Li 2018), CG-SSD applies scaling, rotation, and flipping on individual objects and database sampling for every frame data. Firstly, to keep a more balanced number of categories in the training frame, CG-SSD randomly samples 1, 4, 3, 2, and 2 objects' point clouds for car, bus, truck, pedestrian, and cyclist respectively from the all-training ground truth 3D bounding boxes' point clouds and past them into the current frame. Then, the random flipping along x, y-axis, random rotation of $\left[-\frac{\pi}{4}, \frac{\pi}{4}\right]$ and scaling of [0.95, 1.05] are respectively applied to the objects' boxes and frame's point clouds. In the experiments of verifying that CG-SSD's auxiliary module can be made a plug-in in other models, CG-SSD follows the original model's data augmentation and does not change anything.

Optimization. CG-SSD trains the model with batch size = 30 for 80 epochs on five Nvidia RTX 3090 GPUs in ONCE dataset. The inference speed is measured on an Intel Core i9-10900X CPU and a single Nvidia RTX 3090 GPU. The AdamW (Loshchilov and Hutter 2019) optimizer with a one-cycle policy is used during the training of CG-SSD model. The max learning rate is set to 0.0028, the division factor is set to 10, the momentum range is set from 0.95 to 0.85, and the weight decay is set to 0.01.

Evaluation. For ONCE and Waymo datasets, the 3D IoU-based metric average precision is used for evaluating different models. The IoU between predicted boxes B_p and ground truth boxes B_{gt} can be calculated using the following metric:

$$\text{IoU} = \frac{area\left(B_{gt} \cap B_p\right)}{area\left(B_{gt} \cup B_p\right)}. \tag{6.12}$$

In ONCE dataset, it uses the IoU threshold (0.7, 0.7, 0.7, 0.3, 0.5 for car, bus, truck, pedestrian, and cyclist) to determine whether the predicted box is positive. The predicted orientation angle is added to filter the positive boxes if the angle can fall into the $\pm 90^{\circ}$ range of the matched ground-truth box. The official metric calculates the AP@50 by 50 score thresholds with the recall rates from 0.02 to 1.00 at step 0.02. And it merges the car, bus, and truck into a super-class named vehicle. The mean AP (mAP) is reported by the official metric which is obtained by averaging the AP of three classes (vehicle, pedestrian, and cyclist).

In Waymo dataset, it uses the same method as the ONCE to evaluate the model's performance. The official metric of Waymo dataset provides AP and APH scores. For APH metric, it is calculated by weighting the orientation angle accuracy to the AP. And the orientation angle accuracy is defined as $\min\left(\left|\tilde{\theta}-\theta\right|, 2\pi-\left|\tilde{\theta}-\theta\right|\right)/\pi$, where θ and $\tilde{\theta}$ are the angle of the ground-truth box and predicted box. The IoU threshold of Waymo dataset is set to 0.7, 0.5, and 0.5 for the vehicle, pedestrian, and cyclist views respectively. Additionally, the performance of two difficulty levels is also calculated by the official evaluation metric. "LEVEL 1" boxes have more than five LiDAR points, and "LEVEL 2" have at least one.

6.2.3 Results and Analysis

6.2.3.1 Evaluation on ONCE Testing Set

Table 6.2 shows the result of the CG-SSD module on the testing set of ONCE dataset. The CG-SSD module achieves state-of-the-art results on mAP by only using one frame LiDAR point cloud. On ONCE testing set, the CG-SSD module achieves 68.00% AP, 52.81% AP, and 67.50% AP for the vehicle, pedestrian, and cyclist categories, and 62.77% mAP for three categories. It outperforms all models in the benchmark by at least 1.5% mAP. The inference speed of CG-SSD is 76ms per frame (13FPS), which is slightly slower than CenterPoints (Yin et al., 2021) (67ms, 15FPS). However, CG-SSD gets +1.53% mAP gain by scarifying only 10ms.

Compared with the anchor-free model (CenterPoints (Yin, Zhou, and Krähenbühl 2021)), CG-SSD gets +1.53% mAP better than it. Compared with the anchor-based models (PointPillars (Lang et al. 2019), SECOND (Yan, Mao, and Li 2018), PV-RCNN (Shi et al. 2020)), CG-SSD has better performance on the pedestrian and cyclists with at least +26.72% AP and +5.57% AP. The anchor-based models have better performance (+8.98% AP) in vehicle detection overall anchor-free models, while their performance drops significantly in pedestrian (−30.15% AP) and cyclist (−5.57% AP) detection.

The main reason that leads to this phenomenon, is the limited size of the BEV grids. The pedestrian and cyclist can only occupy one or two pixels in the BEV feature map. So, it is difficult for the network to predict its dimension and orientation angle. Benefitting from the advantages of the auxiliary module for partly visible corner detection, the CG-SSD module shows a larger margin than anchor-based models, especially in the pedestrian and cyclist categories.

6.2.3.2 Evaluation on ONCE Validation Set

On the ONCE benchmark server, CG-SSD can only get the AP and mAP of the three categories. To analyze the CG-SSD's performance at the different distances of an object to the sensor, CG-SSD also evaluates its model on the ONCE validation set. As shown in Table 6.3, the detection precision of the CG-SSD module outperforms CenterPoints (Yin, Zhou, and Krähenbühl 2021) for all objects in different distance ranges. This chapter visualizes the detection result of CenterPoints (Yin, Zhou, and Krähenbühl 2021) and the CG-SSD model in Figure 6.4.

For vehicle detection, CG-SSD has a slight advantage (+0.12% AP) over CenterPoints (Yin, Zhou, and Krähenbühl 2021) in $0-30$m range, and has a significant preference (+1.68% AP & +1.38% AP) in $30-50$m and 50m$-inf$ range. The anchor-based models(PointPillars (Lang et al. 2019), SECOND (Yan, Mao, and Li 2018), PV-RCNN (Shi et al. 2020)) show better performance than CG-SSD (+0.97% AP). The anchor-free model needs to directly predict the object from the incomplete object point clouds, so it is difficult to accurately estimate the dimension and orientation value of the target. However, the anchor-based models regress the difference between the designed anchor and the ground truth box. Even if the predicted value is not accurate enough, the anchor-based models can get results closer to ground truth boxes than the anchor-free models.

TABLE 6.2
Performance Comparison on the ONCE Testing Set

		AP@50(%)				
Model	**Anchor-free**	**Vehicle**	**Pedestrian**	**Cyclist**	**mAP (%)**	**Speed (ms)**
PointPillars* (Lang et al. 2019)	No	69.52	17.28	49.63	45.47	**38**
SECOND* (Yan, Mao, and Li 2018)	No	69.71	26.09	59.92	51.90	40
PV-RCNN* (Shi et al. 2020)	No	**76.98**	22.66	61.93	53.85	167
PointRCNN* (Shi, Wang, and Li 2019)	Yes	52.00	8.73	34.02	31.58	730
CenterPoints* (Yin, Zhou, and Krähenbühl 2021)	Yes	66.35	51.80	65.57	61.24	67
CG-SSD	Yes	68.00	**52.81**	**67.50**	**62.77**	76

Note: * represents the result of the model is from ONCE Benchmark.

TABLE 6.3

Performance Comparison on the ONCE Validation Set

Model	Vehicle (AP@50)(%)				Pedestrian (AP@50)(%)				Cyclist (AP@50)(%)				mAP
	Overall	0–30m	30–50m	50m-inf	overall	0–30m	30–50m	50m-inf	overall	0–30m	30–50m	50m-inf	(%)
PointRCNN* (Shi, Wang, and Li 2019)	52.09	74.45	40.89	16.81	4.28	6.17	2.40	0.91	29.84	46.03	20.94	5.46	28.74
PointPillars* (Lang et al. 2019)	68.57	80.86	62.07	47.04	17.63	19.74	15.15	10.23	46.81	58.33	40.32	25.86	44.34
SECOND* (Yan, Mao, and Li 2018)	71.19	84.04	63.02	47.25	26.44	29.33	24.05	18.05	58.04	69.96	52.43	34.61	51.89
PV-RCNN* (Shi et al. 2020)	**77.77**	**89.39**	**72.55**	**58.64**	23.50	25.61	22.84	17.27	59.37	71.66	52.58	36.17	53.55
CenterPoints* (Yin, Zhou, and Krähenbühl 2021)	66.79	80.10	59.55	43.39	49.90	56.24	42.61	26.27	63.45	74.28	57.94	41.48	60.05
CG-SSD	67.60	80.22	61.23	44.77	**51.50**	**58.72**	**43.36**	**27.76**	**65.79**	**76.27**	**60.84**	**43.35**	**61.63**

Note: * represents the result of the model is from ONCE Benchmark

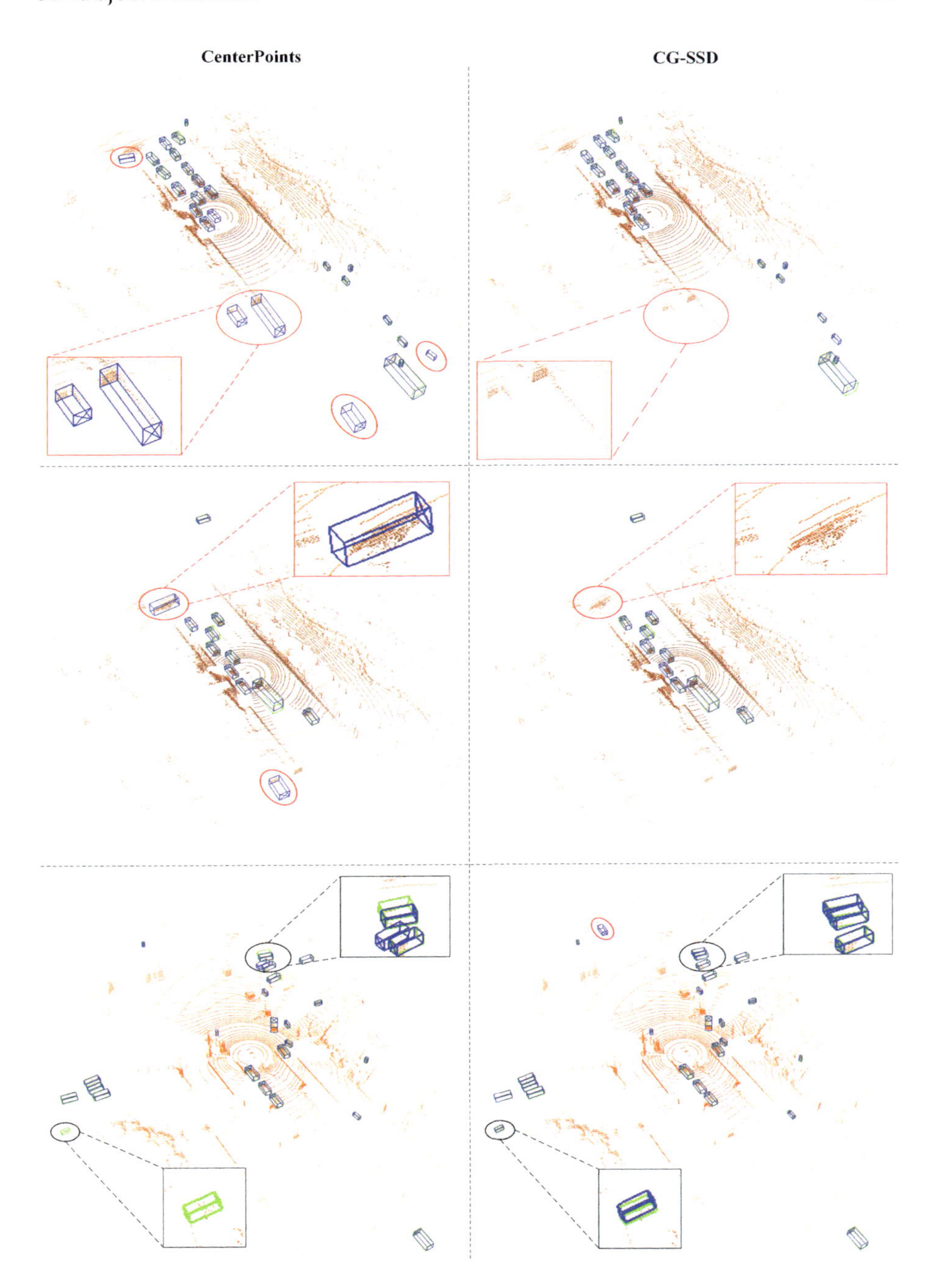

FIGURE 6.4 Visualization of object detection in ONCE. Left: CenterPoints (Yin, Zhou, and Krähenbühl 2021), Right: CG-SSD. The ground truth and predicted bounding boxes are shown in green and blue respectively, the red ovals represent the falsely detected objects and the black ovals are the objects only detected by CG-SSD (Ma et al. 2022).

And for pedestrian and cyclist detection in 50m−*inf*, CG-SSD can get at least +1.39% AP than CenterPoints (Yin, Zhou, and Krähenbühl 2021). The results illustrate that as the distance of the object from the sensor increases, CG-SSD is more stable and more accurate in detecting vehicles than the CenterPoints (Yin, Zhou, and Krähenbühl 2021) model. This is mainly because as the distance increases, the number of points of the target decreases, and the surface contour of the target becomes more blurred. While the center-based model gets the dimension and orientation of the object by aggregating edge features to the center point, it is difficult to learn the accurate 3D information of the object with limited points. However, in the CG-SSD model, it learns partly visible and invisible corners' point information and then fuses corners' features into the object detection task as a way for the model to perceive more accurate 3D information.

6.2.4 Ablation Study

6.2.4.1 Type of Corners

The 3D object on BEV is represented using a rectangle with a rotation angle and has four corners. CG-SSD directly predicts the coordinates of the center point on the vertical direction and the height of the object. So, the 3D bounding boxes can be recovered based on the rectangular box on the BEV and the predicted height. As shown in Figure 6.2, each object's four corners are divided into three types: VC, PVCW, PVCL and IVC. This chapter analyzes the impact of different types of corner detection on ONCE validation set with the CenterPoints (Yin, Zhou, and Krähenbühl 2021) model as the baseline.

As shown in Table 6.4, when CGAM selects the VC and PVCW as the supervision signal, CG-SSD gets a similar performance to the baseline (+0.00% mAP, -0.04% mAP). But when the PVCL and IVC are selected, the model gets significant improvement with baseline (+0.71% mAP, +0.89% mAP). The PVCL and IVC are farther away from the VC and more difficult to be observed by the LiDAR. This result shows that corners that are harder to observe by the sensor can guide the model to learn a more accurate 3D bounding box.

6.2.4.2 Number of Corners

The ablation study of the type of corners shows that different corners contribute differently to the model detection accuracy. To get better results, this chapter tests the CG-SSD model with the different number of corners in the auxiliary module. CenterPoints (Yin, Zhou, and Krähenbühl 2021) is also selected as the baseline. As shown in Table 6.5, when one corner (IVC) is selected as the supervision signal of the CGAM, CG-SSD shows better capacity than the baseline with +0.89% mAP. When two corners (IVC and PVCL) are added to CGAM, the results of the model are improved compared to when one corner is added (+0.18% mAP). When all corners (IVC+PVCL+PVCW+VC) are selected, the number of detection objects of the CGAM increases the difficulty of network convergence. Therefore, the model does not achieve higher detection results due to the increase in the number of detected

TABLE 6.4
Effects of Different Corners for Auxiliary Model on ONCE Validation Set

	Vehicle (AP@50)(%)				Pedestrian (AP@50)(%)				Cyclist (AP@50)(%)				
Corner	overall	0–30m	30–50m	50m-inf	overall	0–30m	30–50m	50m-inf	overall	0–30m	30–50 m	50m-inf	mAP(%)
Baseline	66.79	80.10	59.55	43.39	49.90	56.24	42.61	26.27	63.45	74.28	57.94	41.48	60.05
VC	65.47	79.49	58.77	42.34	49.76	**57.88**	40.11	25.39	**64.93**	**76.17**	60.11	42.43	60.05
PVCW	65.60	79.90	58.52	42.37	50.04	57.25	42.27	26.57	64.38	75.65	58.48	42.34	60.01
PVCL	66.86	**80.13**	59.69	42.37	**50.83**	57.86	**42.90**	27.73	64.58	75.34	**60.33**	42.02	60.76
IVC	**67.34**	79.92	**59.71**	**44.48**	50.75	57.85	42.76	**27.89**	64.72	75.79	59.84	**42.82**	**60.94**

TABLE 6.5
Effects of Different Numbers of Corners for Auxiliary Model on ONCE Validation Set

Number of	Vehicle (AP@50)(%)				Pedestrian (AP@50)(%)				Cyclist (AP@50)(%)				
Corners	overall	0–30m	30–50m	50m-inf	overall	0–30m	30–50m	50m-inf	overall	0–30m	30–50m	50m-inf	mAP(%)
Baseline	66.79	80.10	59.55	43.39	49.90	56.24	42.61	26.27	63.45	74.28	57.94	41.48	60.05
One*	67.34	79.92	59.71	44.48	50.75	57.85	42.76	**27.89**	64.72	75.79	59.84	42.82	60.94
Two*	67.17	80.02	60.33	43.99	50.57	57.77	42.63	27.20	65.63	76.25	60.68	42.77	61.12
Three*	67.60	**80.22**	**61.23**	44.77	**51.50**	**58.72**	**43.36**	27.76	**65.79**	76.27	**60.84**	43.35	**61.63**
Four*	**68.53**	79.97	61.18	**46.89**	49.38	56.64	41.46	26.20	65.39	**76.45**	60.35	**44.09**	61.10

Note: One: IVC, Two: IVC + PVCL, Three: IVC + PVCL + PVCW, Four: IVC + PVCL + PVCW + VC

corners. The CG-SSD model gets the best result when three corners (IVC, PVCL, and PVCW) are selected as the detection object (+1.58% mAP vs baseline).

6.2.4.3 Effects of Different Corner Auxiliary Signal

To select the most beneficial supervision signal in the CGAM, this chapter selects the corners classification and position information as the output of the auxiliary module to improve the detection accuracy of the model. As shown in Table 6.6, similar to the previous model (Wang et al. 2020; Hu et al. 2021) results, when only the corners classification information is added, the detection accuracy of the model will be improved by +0.34% mAP compared with CenterPoints (Yin, Zhou, and Krähenbühl 2021). The experiment also shows that adding the corners offset is more beneficial to the detection accuracy of the model with +1.24% mAP. The ablation study proves that adding the classification and regression signals of the corners can guide the network to learn more accurate center point positions and bounding boxes of the target.

6.2.4.4 Extension to BEV-based Models

The CGAM consists of a series of 2D convolutional modules and is easily extended to any 3D detector that uses a BEV map to detect objects. To verify it can play a plug-in to other models, this chapter selects PointPillars (Lang et al. 2019), SECOND (Yan, Mao, and Li 2018), and PV-RCNN (Shi et al. 2020) to test on ONCE and Waymo validation set (based on 20% training data). In Figure 6.5, this chapter visualizes the detection results of SECOND (Yan, Mao, and Li 2018) and enhanced SECOND with CGAM. As can be seen from the figure, the model can get better detection results after adding the CGAM module. As shown in Tables 6.7 and 6.8, on ONCE dataset, the auxiliary module can bring +3.37%~+7.05% mAP to the original detector. On Waymo dataset, enhanced models outperform the original model with +1.31%~+14.23% APH in L1 level and +1.17%~+12.93% APH in L2 level. Especially, for pedestrian and cyclist detection, CG-SSD's corner detection module has more advantages. These results show that after adding the CGAM, all three detectors get better performance on two datasets. Further, it can be extended as a plug-in to most 3D detectors that use a BEV map.

6.2.4.5 Selection of Optimal Hyperparameters

To select the reasonable hyperparameters in the CG-SSD model, this chapter supplies the experiments with the different settings of those parameters as below.

Voxel size. The voxel size is the key parameter for the voxelization of point clouds and it affects the object detection accuracy of the model at different scales. Table 6.9 demonstrates the effects of changing voxel size in CG-SSD. The smaller voxel size is useful for small object detection but will bring more computation and a more obvious "empty hole" phenomenon on the BEV map for large objects. The mAP decreases by 0.54% when using a small voxel size ([0.08m, 0.08m, 0.2m]). The larger voxel size will lead to loss of local information, and may cause adjacent small objects to be assigned to the same pixel and make worse accuracy. And the mAP drops 5.65% when using large voxel size ([0.12m, 0.12m, 0.2m]). For fair comparisons, the voxel size is set to [0.1m, 0.1m, 0.2m] for all the models (e.g. CenterPoints (Yin, Zhou, and Krähenbühl 2021)) in all experiments.

TABLE 6.6

Effects of Different Corner Auxiliary Signal on ONCE Validation Set

Method	Vehicle (AP@50)(%)				Pedestrian (AP@50)(%)				Cyclist (AP@50)				mAP
	overall	0–30m	30–50m	50m-inf	overall	0–30m	30–50m	50m-inf	overall	0–30m	30–50m	50m-inf	(%)
Baseline	66.79	80.10	59.55	43.39	49.90	56.24	42.61	26.27	63.45	74.28	57.94	41.48	60.05
w/ classification	66.65	79.44	59.28	44.10	49.12	56.19	41.90	25.23	65.41	**76.46**	60.31	**44.23**	60.39
w/ classification&offset	**67.60**	**80.22**	**61.23**	**44.77**	**51.50**	**58.72**	**43.36**	**27.76**	**65.79**	76.27	**60.84**	43.35	**61.63**

TABLE 6.7

Extension to Some BEV Map-based Models on ONCE Validation Set

Model	AP@50(%)			mAP (%)
	Vehicle	Pedestrian	Cyclist	
PointPillars* (Lang et al. 2019)	68.57	17.63	46.81	44.34
+auxiliary module	**70.45 (+1.88)**	**18.97 (+1.34)**	**53.71 (+6.9)**	**47.71 (+3.37)**
SECOND* (Yan, Mao, and Li 2018)	**71.19**	26.44	58.04	51.89
+auxiliary module	71.10 **(−0.09)**	**38.2 7 (+11.83)**	**61.89 (+3.85)**	**57.09 (+5.2)**
PV-RCNN* (Shi et al. 2020)	77.77	23.50	59.37	53.55
+auxiliary module	**79.20 (+1.43)**	**37.19 (+13.69)**	**65.39 (+6.02)**	**60.60 (+7.05)**

Note: *represents the result from ONCE dataset paper (Mao et al. 2021).

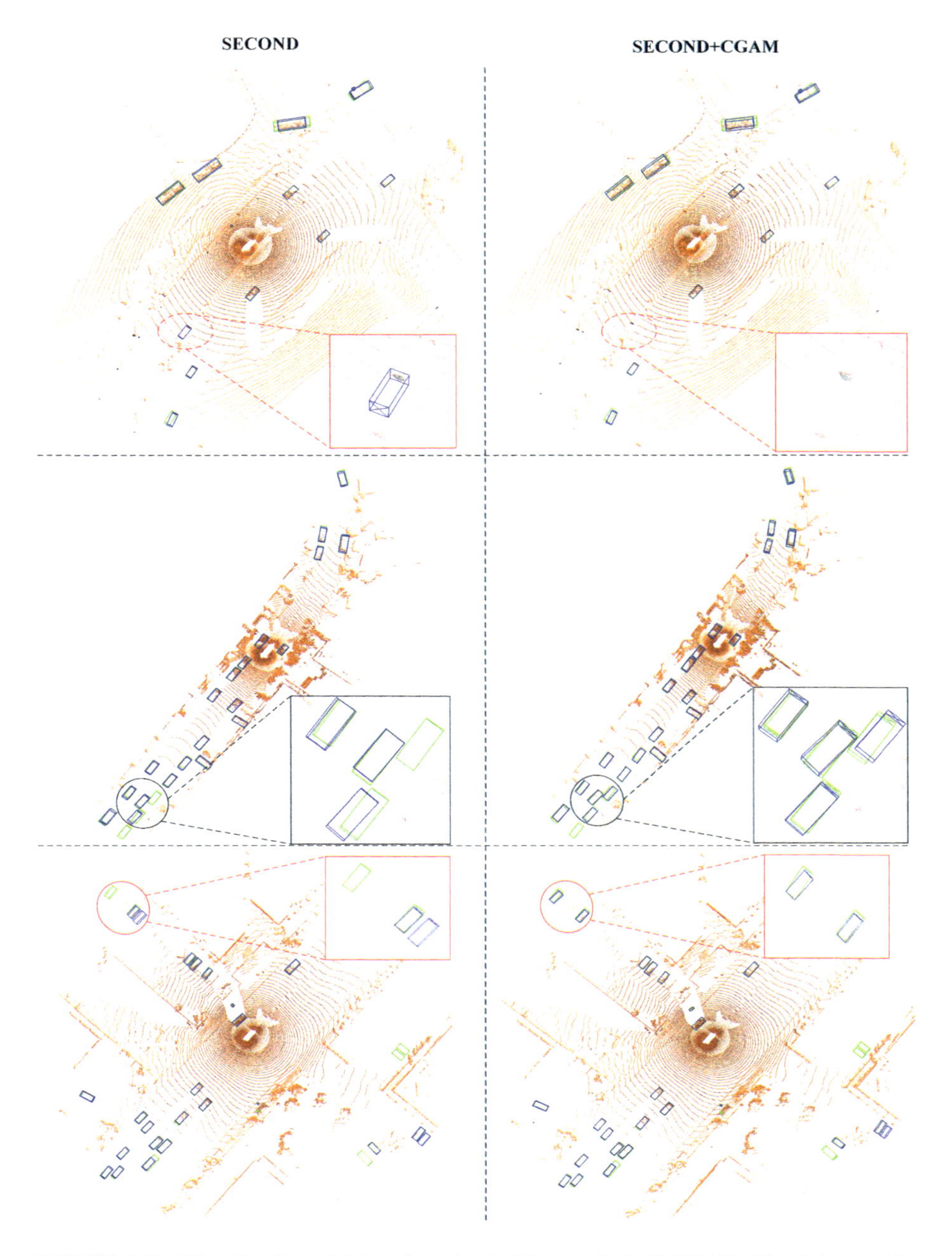

FIGURE 6.5 Visualization of object detection in Waymo. Left: SECOND (Yan, Mao, and Li 2018), Right: SECOND+CGAM. The ground truth and predicted bounding boxes are shown in green and blue respectively, the red ovals represent the falsely detected objects and the black ovals are the objects only detected by SECOND+CGAM. (Ma et al. 2022).

TABLE 6.8

Extension to Some BEV Map-based Models on Waymo Validation Set

Model	Vehicle/APH(%)		Pedestrian/APH(%)		Cyclist/APH(%)	
	L1	L2	L1	L2	L1	L2
PointPillars* (Lang et al. 2019)	69.83	61.64	46.32	40.64	51.75	49.80
+auxiliary module	**71.87 (+2.04)**	**63.93 (+2.29)**	**60.55 (+14.23)**	**53.57 (+12.93)**	**63.86 (+12.11)**	**61.51 (+11.71)**
SECOND* (Yan, Mao, and Li 2018)	70.34	62.02	54.24	47.49	55.62	53.53
+auxiliary module	**71.65 (+1.31)**	**63.78 (+1.76)**	**58.01 (+3.77)**	**51.03 (+3.54)**	**63.31 (+7.69)**	**61.01 (+7.48)**
PV-RCNN* (Shi et al. 2020)	74.74	66.80	61.24	53.95	64.25	61.82
+auxiliary module	**76.53 (+1.79)**	**67.97 (+1.17)**	**66.62 (+5.38)**	**58.36 (+4.41)**	**69.63 (+5.38)**	**67.12 (+5.30)**

Note: * represents the results from the open-source project OpenPCDet.

TABLE 6.9

The Result of CG-SSD Model on ONCE Dataset with Different Voxel Sizes

Voxel Size(m)	Vehicle (AP@50)(%)				Pedestrian (AP@50)(%)				Cyclist (AP@50)(%)				mAP(%)
	overall	0–30m	30–50m	50m-inf	overall	0–30m	30–50m	50m-inf	overall	0–30m	30–50m	50m-inf	
(0.08, 0.08, 0.2)	65.55	**80.56**	58.83	38.57	**52.07**	**60.11**	**44.31**	27.18	65.64	**78.18**	59.42	42.23	61.09
(0.1, 0.1, 0.2)	**67.60**	80.22	**61.23**	**44.77**	51.50	58.72	43.36	**27.76**	**65.79**	76.27	**60.84**	**43.35**	**61.63**
(0.12, 0.12, 0.2)	62.55	76.81	54.22	39.53	43.38	50.80	35.82	21.77	62.03	73.89	56.28	37.81	55.98

Kernel size in 2D backbone. Different kernel sizes in 2D convolution can get multi-receptive fields to learn features on the BEV map. In Table 6.10, this chapter shows the result of the CG-SSD model with different kernel sizes in 2D backbones. The CG-SSD achieves the highest performance with the kernel size set to 3. Small kernel size (e.g. 1) cannot get a sufficient receptive field to learn local neighborhood features for large object detection (e.g. vehicle). The overall AP of vehicle drops 9.77% when using a small kernel size. The large kernel size (e.g. 5) can get more features from the adjacent pixels but lose the local detail features. Thus, the performance drops 3.52% on small object detection (e.g. pedestrian) when choosing a large kernel size.

Radius. The radius is used to calculate the 2D Gaussian value (Equation 3) as the target of the classification branch. The closer the pixel is to the center, the larger its Gaussian value and the greater tl probability of being predicted as the center. Thus, the Gaussian value can reduce the amount of penalty of negative pixels on the BEV map during network training. Table 6.11 demonstrates that a larger radius causes extra pixels farther from the center to be assigned with non-zero Gaussian values resulting in a decrease in the performance (−2.41% mAP). The smaller radius increases the amount of penalty of negative pixels and gets worse accuracy (−0.93% mAP).

Loss weight of CGAM module. The loss weight of CGAM module determines the contribution to total loss and the importance of the corners' information. In order to get the best result, the CG-SSD is trained with different loss weights of the CGAM module. As shown in Table 6.12, CG-SSD can get the best result when the loss weight of the CGAM module is 0.25.

6.3 DISCUSSION

There are a large number of LiDAR points around the visible corner of objects, and it is easily detected by the network. Therefore, adding the learning corners' information to the auxiliary module cannot improve the detection accuracy of the network. When this chapter adds the partly visible corners to the auxiliary module, CG-SSD model gets better results with the selected corner is further away from the visible corner (Section 6.2.4.1). The ablation study of the types of corners shows that CGAM can improve detection accuracy, and learning invisible corner information is more important for the detector.

The object is represented by a rectangle in the BEV map, and the three corners can determine the shape of a rectangle. Therefore, the model can perceive the complete target information and achieve the best detection results when three corners are selected as the supervision signals (Section 6.2.4.2). Selecting four corners as the detection object cannot achieve better accuracy, which will not only generate redundant information but also increase the complexity and computational burden of the model. Furthermore, selecting the corner effective supervision signal is also a critical problem in the CGAM. The corner classification and offset value are the key features and they can bring a significant improvement in detection accuracy (Section 6.2.4.3). The precise location information of the corners can play an important guiding role when the network predicts the bounding boxes.

TABLE 6.10

The Result of CG-SSD Model on ONCE Dataset with Different Kernel Sizes in 2D Backbone

	Vehicle (AP@50)(%)				Pedestrian (AP@50)(%)				Cyclist (AP@50)(%)				
Kernel Size	overall	0–30m	30–50m	50m-inf	overall	0–30m	30–50m	50m-inf	overall	0–30m	30–50m	50m-inf	mAP(%)
1	57.83	74.69	46.93	31.90	43.91	49.11	38.73	23.73	58.81	71.74	52.56	33.36	53.52
3	**67.60**	**80.22**	**61.23**	**44.77**	**51.50**	**58.72**	**43.36**	**27.76**	**65.79**	**76.27**	**60.84**	**43.35**	**61.63**
5	66.42	79.59	58.42	43.28	47.98	53.59	42.13	26.24	64.16	74.41	59.97	40.91	59.52

TABLE 6.11

The Result of CG-SSD Model on ONCE Dataset with Different Radius

	Vehicle (AP@50)(%)				Pedestrian (AP@50)(%)				Cyclist (AP@50)(%)				
Radius	Overall	0–30m	30–50m	50m-inf	Overall	0–30m	30–50m	50m-inf	Overall	0–30m	30–50m	50m-inf	mAP(%)
1	67.08	**80.27**	59.83	43.61	50.99	58.67	42.65	25.23	64.03	74.51	59.69	**43.76**	60.70
2	**67.60**	80.22	**61.23**	**44.77**	**51.50**	**58.72**	**43.36**	**27.76**	**65.79**	**76.27**	**60.84**	43.35	**61.63**
4	64.69	77.77	59.06	42.67	48.88	54.52	42.71	27.41	64.08	74.19	59.79	41.45	59.22

TABLE 6.12

The Result of CG-SSD Model on ONCE Dataset with Different Loss Weights for CGAM Module

	Vehicle (AP@50)(%)				Pedestrian (AP@50)(%)				Cyclist (AP@50)(%)				
Weight	Overall	0–30m	30–50m	50m-inf	Overall	0–30m	30–50m	50m-inf	Overall	0–30m	30–50m	50m-inf	mAP(%)
0.12	67.49	80.08	**61.26**	44.28	50.63	57.53	**43.95**	26.12	64.29	74.20	59.79	42.66	60.80
0.25	**67.60**	**80.22**	61.23	**44.77**	**51.50**	**58.72**	43.36	**27.76**	**65.79**	76.27	**60.84**	**43.35**	**61.63**
0.50	64.87	79.00	57.98	40.59	48.64	55.79	40.85	24.89	65.05	**76.43**	60.17	42.45	55.98

The CGAM consists of a series of CBR blocks, and it is easy to be extended to other models that use the BEV feature map to predict objects. The CGAM is added to three anchor-based models. Due to the CGAM providing more information of the object corners and guiding the network to predict more accurate 3D bounding boxes, the improved models get better performance than the original models on ONCE and Waymo datasets (Section 6.2.4.4). Especially, because the CGAM provides different corners' locations that can implicitly express the position of the center point, the improved anchor-based models show better capacity for small objects (pedestrians and cyclists).

For vehicle detection, the CG-SSD model result is lower than anchor-based models. Different from 2D objects in images containing dense pixels rounding their central location, 3D objects have incomplete point clouds. And there are no points at their physical 3D center location. After projecting the sparse voxel features to the BEV map, the large-scale objects (e.g. vehicles) generally have fewer features learned from 3D backbones and create "empty holes" rounding their central location. Therefore, it is difficult for anchor-free models to directly predict the accurate center and size of large objects. For small objects (e.g. pedestrians and cyclists), their 3D bounding boxes have a similar size to the grid size of the BEV map (e.g. the mean 3D size of the pedestrian is [0.76 m, 0.75 m, 1.69 m] and the grid size is [0.8 m, 0.8 m]), and will not create an "empty hole" at the center of the object on the projected BEV feature map. The CG-SSD model can directly regress the accurate center and dimension of small objects. Besides, the corner detection auxiliary module proposed by CG-SSD model can provide accurate corner information of the target, which greatly reduces the losses of target information and thus improves the detection accuracy. For pedestrian and cyclist detection, the CG-SSD model shows more significant improvement compared with other models.

6.4 CONCLUSION

Due to the placement position of the LiDAR sensor, the scanning angle, and the occlusion between the targets, the LiDAR sensor cannot obtain complete target point cloud data. It is very difficult to predict and regress accurate 3D bounding boxes of objects from incomplete object point clouds. This chapter introduced a one-stage corner-guided anchor-free 3D object detection model (CG-SSD). The CG-SSD model innovatively takes the corner information of the object into the training and inference of the network, to guide the network to obtain more accurate 3D bounding boxes from the limited point clouds. An adaptive corner classification method is designed according to the limited number of object points. Different from the previous model to obtain the 3D bounding box of the target by detecting the center point, CG-SSD innovatively added the corner prediction module CGAM. The experiment on the ONCE dataset shows that the CG-SSD model achieves better performance than other state-of-the-art models (CenterPoints (Yin, Zhou, and Krähenbühl 2021) and PV-RCNN (Shi et al. 2020)) for supervised 3D object detection using single frame point clouds data. In addition, the results on ONCE and Waymo data show that CGAM can be extended to other 3D detectors that use the BEV map as a plug-in for boosting the performances. The results demonstrate the importance of corner detection in 3D object detection. In future work, this chapter will focus on building a corner point-based model instead of center-based ones to fuse different sensor modalities to detect and track 3D objects continuously in space and time for fully self-driving cars.

NOTE

1 https://github.com/open-mmlab/OpenPCDet.

REFERENCES

Dai, Angela, Christian Diller, and Matthias Niebner. 2020. "SG-NN: Sparse Generative Neural Networks for Self-Supervised Scene Completion of RGB-D Scans." In *Proceedings of the IEEE Computer Society Conference on Computer Vision and Pattern Recognition*, 846–55. doi:10.1109/CVPR42600.2020.00093

Geiger, A, P Lenz, C Stiller, and R Urtasun. 2013. "Vision Meets Robotics: The KITTI Dataset. The International Journal of Robotics Research." *The International Journal of Robotics Research*, no. October: 1–6.

Hu, Yihan, Zhuangzhuang Ding, Runzhou Ge, Wenxin Shao, Li Huang, Kun Li, and Qiang Liu. 2021. "AFDetV2: Rethinking the Necessity of the Second Stage for Object Detection from Point Clouds." *arXiv* 2112.09205 (December).

Lang, Alex H., Sourabh Vora, Holger Caesar, Lubing Zhou, Jiong Yang, and Oscar Beijbom. 2019. "Pointpillars: Fast Encoders for Object Detection from Point Clouds." In *Proceedings of the IEEE Computer Society Conference on Computer Vision and Pattern Recognition*, 12689–97. doi:10.1109/CVPR.2019.01298

Law, Hei, and Jia Deng. 2020. "CornerNet: Detecting Objects as Paired Keypoints." *International Journal of Computer Vision* 128 (3): 642–56. doi:10.1007/s11263-019-01204-1

Liu, Wei, Dragomir Anguelov, Dumitru Erhan, Christian Szegedy, Scott Reed, Cheng Yang Fu, and Alexander C. Berg. 2016. "SSD: Single Shot Multibox Detector." *arXiv* 1512.02325 (December). doi:10.1007/978-3-319-46448-0_2

Loshchilov, Ilya, and Frank Hutter. 2019. "Decoupled Weight Decay Regularization." *arXiv* 1711.05101 (November).

Ma, Ruiqi, Chi Chen, Bisheng Yang, Deren Li, Haiping Wang, Yangzi Cong, and Zongtian Hu. 2022. "CG-SSD: Corner Guided Single Stage 3D Object Detection from LiDAR Point Cloud." *ISPRS Journal of Photogrammetry and Remote Sensing* 191: 33–48. doi:10.1016/j.isprsjprs.2022.07.006

Mao, Jiageng, Minzhe Niu, Chenhan Jiang, Hanxue Liang, Jingheng Chen, Xiaodan Liang, Yamin Li, et al. 2021. "One Million Scenes for Autonomous Driving: ONCE Dataset." *arXiv* 2106.11037.

Shi, Shaoshuai, Chaoxu Guo, Li Jiang, Zhe Wang, Jianping Shi, Xiaogang Wang, and Hongsheng Li. 2020. "PV-RCNN: Point-Voxel Feature Set Abstraction for 3D Object Detection." In *Proceedings of the IEEE Computer Society Conference on Computer Vision and Pattern Recognition*, 10526–35. doi:10.1109/CVPR42600.2020.01054

Shi, Shaoshuai, Xiaogang Wang, and Hongsheng Li. 2019. "PointRCNN: 3D Object Proposal Generation and Detection from Point Cloud." In *Proceedings of the IEEE Computer Society Conference on Computer Vision and Pattern Recognition*, 770–79. doi:10.1109/CVPR.2019.00086

Sun, Pei, Henrik Kretzschmar, Xerxes Dotiwalla, Aurélien Chouard, Vijaysai Patnaik, Paul Tsui, James Guo, et al. 2020. "Scalability in Perception for Autonomous Driving: Waymo Open Dataset." In *Proceedings of the IEEE Computer Society Conference on Computer Vision and Pattern Recognition*, 2443–51. doi:10.1109/CVPR42600.2020.00252

Wang, Guojun, Bin Tian, Yunfeng Ai, Tong Xu, Long Chen, and Dongpu Cao. 2020. "CenterNet3D:An Anchor Free Object Detector for Autonomous Driving." *arXiv* 2007.07214 (July).

Yan, Yan, Yuxing Mao, and Bo Li. 2018. "Second: Sparsely Embedded Convolutional Detection." *Sensors (Switzerland)* 18 (10): 1–17. doi:10.3390/s18103337

Yin, Tianwei, Xingyi Zhou, and Philipp Krähenbühl. 2021. "Center-Based 3D Object Detection and Tracking." In *Proceedings of the IEEE Computer Society Conference on Computer Vision and Pattern Recognition*, 11779–88. doi:10.1109/CVPR46437.2021.01161

7 Point Cloud Semantic Segmentation

7.1 METHODOLOGY

7.1.1 Brief Description of the Methodology

The objective of point cloud semantic segmentation tasks is to provide per-point semantic information for a given scene, segmenting point clouds of different categories in real-world scenarios (such as lampposts, roads, buildings, trees, etc.). This involves methods of projection representation, discretization representation, and unstructured representation. Existing methods generally perform well in indoor scene point clouds and outdoor single-frame point clouds but struggle to be directly applied to the segmentation tasks of large-scale outdoor scene point clouds (e.g., MLS) due to not considering the following characteristics of large-scale outdoor point clouds:

1. Severe sample imbalance. Generally, the ground and trees make up the majority, while some are in the minority (such as pedestrians, traffic lights, signs). Thus, when using deep neural networks for feature learning, the learning process tends to be dominated by the majority class (e.g., ground, trees), with the minority class being overlooked, making it difficult for the minority samples to learn;
2. Large data scale. MLS point cloud data of road scenes may contain tens of millions of points, and the computational load is significant when these points are directly processed by the network;
3. Significant variation in object scales. In urban scenes, small targets may only be one meter in length, while large targets can extend to hundreds of meters, meaning that the selection of local areas for feature learning can significantly impact network performance;
4. Complex and diverse structures. Many outdoor objects are highly complex and varied (e.g., buildings) and may overlap with each other (e.g., lampposts and trees).

Addressing the issues mentioned above, this chapter accomplishes the following tasks:

1. Given the massive quantity of points and the varying sizes of different categories, this study adopts a hierarchical sampling strategy, including Uniform Sampling (US), Random Sampling (RS), and Farthest Point Sampling (FPS), to prevent neglecting the sparse distribution of point clouds and small-sized targets when applying spatial downsampling. This ensures that

DOI: 10.1201/9781003486060-10

objects remain covered even at higher levels, making the approach simple yet effective.

2. Considering the complex spatial relationships within MLS point clouds and the characteristics of point clouds, such as "absolute coordinates and size" and "irregular distribution", this chapter designs a point-based local feature extraction module that integrates relative spatial information with feature depth information.
3. To address the learning challenges of minority samples due to sample imbalance, a specially designed loss function is employed, enabling the network to focus more on the learning of minority samples during the training process.

The method presented in this chapter directly uses point cloud coordinates (*xyz*) as input and predicts the semantic labels for each point in an end-to-end manner. The overall framework is illustrated in Figure 7.1. The proposed network employs a U-shaped structure with a downsampling-upsampling architecture and cross-layer connections as the backbone network. The downsampling process maps features to a higher feature space layer-by-layer, expanding the receptive field to obtain higher-level, more abstract features. The sampling process aims to progressively propagate abstract features to each point, thereby achieving per-point prediction. Feature abstraction plays a crucial role in this method, and we propose a network structure for feature aggregation and an efficient strategy for spatial downsampling. The downsampling process includes: (1) Spatial Downsampling; (2) Point Grouping; (3) Arbitrary Position Activation; (4) Local Feature Aggregation.

7.1.2 Spatial Downsampling

To ensure computational efficiency while addressing small-scale and low-density targets, the sampling process should be streamlined. This chapter divides the spatial sampling of point clouds into three parts: Uniform Sampling (US), Random Sampling (RS), and Farthest Point Sampling (FPS), each employed in different network layers. All sampling processes operate within coordinate space.

Initially, the entire scene is sampled using a uniform grid to alleviate the uneven point distribution of MLS point clouds to some extent. Grid sampling is used instead of density-based uniform sampling to divide the scene into grids based on a specific resolution of sampling. A lower resolution should be used here, as a higher sampling density is not conducive to learning small-scale targets. Subsequently, random sampling is utilized in the first few layers of the network, as the efficiency of FPS is greatly affected by the number of input points, leading to high computational complexity in the early layers, and FPS is easily influenced by noise points. FPS is used for downsampling in the last few layers because the point cloud is relatively sparse at this stage, and random sampling points cannot effectively cover targets with lower density. Moreover, due to the reduced number of points in the later layers, using FPS does not significantly impact efficiency.

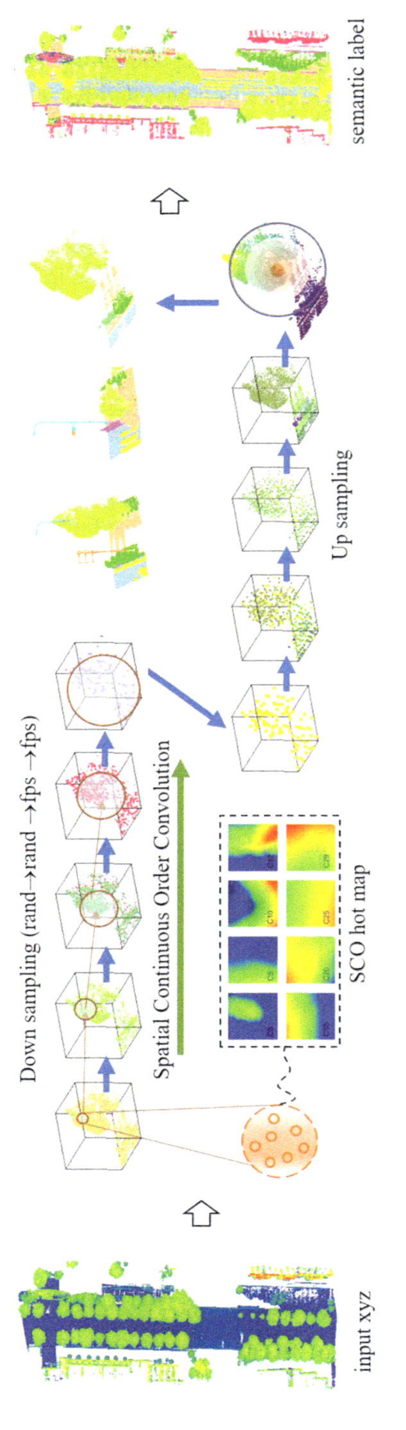

FIGURE 7.1 The framework of the proposed point cloud semantic segmentation method.

7.1.3 Local Features Aggregation

Given the input point set $X^{\text{in}} \in \mathbb{R}^{N\times 3}$ and corresponding input features $F^{\text{in}} \in \mathbb{R}^{N\times D}{}_{\text{in}}$, the objective is to model the local spatial representation of the downsampled point set $X \in \mathbb{R}^{S\times 3}$. The aggregated feature of the point cloud is represented as $F \in \mathbb{R}^{S\times D'}$. For simplicity, this section focuses on local feature aggregation at one position, though this method can be applied to the entire point set for feature abstraction. Here, x denotes one position in the point set and f denotes the corresponding features. Due to the permutation-invariant property of point clouds, the discussed functions are independent of the input point order. The feature aggregation process is illustrated in Figure 7.2.

7.1.3.1 Points Grouping

Prior to each aggregation process, the nearby points of each downsampled point are initially grouped as the local unit for aggregation, denoted as $\mathcal{N}(x)$, utilizing the k-nearest neighbors (KNN) method. Specifically, a larger $K = \sigma k$ is set, where σ represents a dilation ratio greater than one (e.g., $\sigma = 1.5$) for KNN calculation. Subsequently, k points are randomly sampled in the first few layers to slightly expand the receptive field. However, it's important to note that a larger σ may lead to poorly represented positions for x. It should be noted that the query neighbors are identified in the point set X^{in} prior to downsampling. To encode the input features, the fused points encompass the original spatial position, relative spatial position, and relative features (edge features). For each position x, the original spatial information of its ith neighbor is defined as follows:

$$s_i^{ab} = x \oplus x_i \oplus x^{gl} \oplus x_i^{gl} \tag{7.1}$$

Here, $x_i \in \mathcal{N}(x)$, and x^{gl} represents the global position before partitioning into blocks. The symbol $\oplus$ denotes concatenation. It is important to note that $x \in X$ and $x_i \in X^{in}$. The relative spatial information is defined as follows:

$$s_i^{rl} = (x - x_i) \oplus \| x - x_i \| \tag{7.2}$$

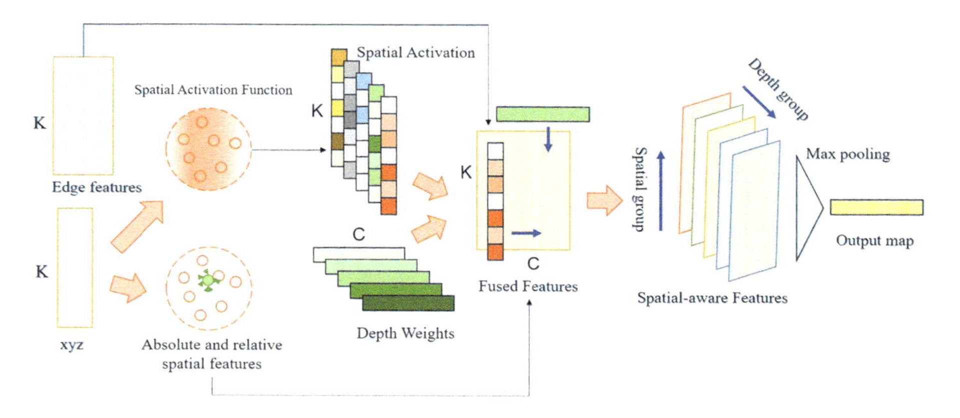

FIGURE 7.2 Feature aggregation module. This module mainly contains points grouping, arbitrary position mapping, spatial and channel direction aggregating.

where $\|\cdot\|$ denotes the calculation of the Euclidean distance. The edge feature is as follows:

$$f_i^{rl} = f^{in} - f_i^{in} \tag{7.3}$$

The features are concatenated straightforwardly, resulting in the fused features of the ith neighbor of position x is:

$$f_i = s_i^{ab} \oplus s_i^{rl} \oplus f_i^{rl} \tag{7.4}$$

Subsequently, $f \in \mathbb{R}^{k \times D}$ is obtained, representing the fused feature of position x.

7.1.3.2 Arbitrary Spatial Position Mapping

Just as convolutional neural networks excel in learning spatial features for image processing tasks, 3D convolution is similarly employed for spatial feature learning in three-dimensional data. The convolution function with kernel g is defined as follows:

$$(\mathcal{F} * \mathcal{G})(x) = \iint_{\tau \in \mathbb{R}^d} \mathcal{F}(\tau)\mathcal{G}(x+\tau)d\tau \tag{7.5}$$

The convolution kernel g is defined by kernel function $\mathcal{G}$. Unlike other convolutional methods, such as those proposed by (Xu et al. 2018), (Wu, Qi, and Fuxin 2019) and (Hermosilla et al. 2018), which aim to approximate the continuous kernel using Monte Carlo estimation, the proposed method directly learns a function to map an arbitrary position in continuous Euclidean space to its corresponding activation value. In simpler terms, when a spatial position x_i is input, relative to the local coordinate system with x as the origin, the activation value of the corresponding position is obtained.

$$g_i = \mathcal{G}(x - x_i) \tag{7.6}$$

The objective is to train the kernel function and utilize it for predicting activation values during inference. It's important to distinguish between g and $\mathcal{G}$: different features at the same position share the same g, while different positions share the same $\mathcal{G}$. Here, $\mathcal{G}$ is constructed with multiple layers of 1×1 convolutions, batch normalization, and non-linear activation. In scenarios where points are uniformly sampled, g is consistent across different positions in X, similar to images.

7.1.3.3 Local Feature Aggregating

To obtain the features of $N(x)$, aggregation is performed in both spatial and channel directions:

$$\breve{f} = \mathcal{S} \circ \mathcal{D}(f \mid w, g) \tag{7.7}$$

Here, $\mathcal{S}$ represents spatial direction aggregation, while $\mathcal{D}$ signifies channel direction aggregation. $w \in \mathbb{R}^{D'}$ denotes learnable weights. $g \in \mathbb{R}^{k}$ represents spatial activation values, and $\check{f} \in \mathbb{R}^{D'}$ denotes the output features of position x.

$$\hat{f}_i^{(d')} = \sigma\left(\sum_{d=1}^{D} f_i^{(d)} w^{(d)(d')}\right) \tag{7.8}$$

$\hat{f}^{(d)}$ represents the dth feature of ith neighbor of position x. $w^{(d)(d')}$ signifies the dth value of the d'^{th} kernel w. σ is a non-linear activation function.

This process is iterated multiple times to extract higher-level features. Subsequently, function $\mathcal{G}$ is employed to compute the spatial activation value g of position x. As the activation kernel traverses through the feature channel direction, each position corresponds to only one kernel value. This kernel is utilized for spatial direction encoding.

$$\check{f}^{(d')} = max\left\{\sigma\left(g \cdot \hat{f}^{(d')}\right)\right\} \tag{7.9}$$

The symbol $\cdot$ denotes element-wise product, while *max* represents the max pooling operation, which condenses the vector to a single value.

These functions aggregate the features in both spatial and channel directions. The distinction between g and w lies in g being contingent upon spatial position, whereas w corresponds to feature channels. Despite seeming to serve different purposes, the order of operations can impact results, signifying that $\mathcal{S} \circ \mathcal{D} \neq \mathcal{D} \circ \mathcal{S}$ because spatial aggregation precedes channel aggregation. This order leverages max pooling for spatial aggregation to identify the most representative features across positions, followed by channel aggregation. However, all aforementioned methods can be easily implemented using matrix multiplication.

7.1.4 Feature Abstraction and Propagation

7.1.4.1 Feature Abstraction

As mentioned in Section 7.1.3, the aggregation function plays a crucial role in expanding the receptive field by consolidating the local features of the point set X. Consequently, multiple similar aggregation blocks are stacked to progressively elevate the features to a higher level, incorporating more abstract information, until the receptive field encompasses most of the space. Prior to each aggregation process, various spatial sampling methods are employed to decrease the number of points based on the level of abstraction. Subsequently, both the sampled and unsampled points are utilized for feature aggregation.

$$F^{l} = \mathcal{A}\left(X^{l-1}, X^{l}, F^{l-1}\right) \quad l = \{0, \ldots, L\} \tag{7.10}$$

$\mathcal{A}$ is an aggregation function primarily composed of symmetric functions $\mathcal{G}$, $\mathcal{S}$ and $\mathcal{D}$, all of which ensure permutation invariance:

$$\mathcal{A}\left(X_{\mu}^{l-1}, X_{\mu}^{l}, F_{\mu}^{l-1}\right) = \mathcal{A}\left(X_{\nu}^{l-1}, X_{\nu}^{l}, F_{\nu}^{l-1}\right) \tag{7.11}$$

where μ and ν represent two different orders.

7.1.4.2 Feature Propagation

While deconvolution is commonly employed in the upsampling process, this section does not propose a corresponding method for point aggregation, assuming that appropriate features have been acquired during aggregation. Here's the process:

1. Initially, to obtain the lth layer points, upsampling utilizes the trilinear interpolation method, akin to (Qi et al. 2017), by using the lth-layer coordinates to achieve weighted-sum features as interpolated points.
2. Subsequently, the interpolated points are concatenated with the points from the corresponding layer $\tilde{l}$ during the aggregation process using skip connections.
3. Lastly, the fused points are projected into a lower-dimensional feature space. The propagation process can be defined as follows:

$$P^{l} = \mathcal{M}\left(\mathcal{I}\left(X^{l-1}, X^{\tilde{l}}, F^{l-1}\right) \oplus F^{\tilde{l}}\right) \quad \tilde{l} = \{L, \ldots, 0\} \tag{7.12}$$

where $l = 2L - \tilde{l}$ and $\mathcal{M}$ represents feature mapping such as MLP, $\mathcal{I}$ denotes the interpolation function. This process is iterated multiple times until the number of upsampled points equals the number of input points.

7.1.5 Loss Function

Given the significant differences in the number of points across different categories within the MLS point cloud dataset, while it is possible to simply assign different class weights to balance the network's supervisory signal, different weight allocation strategies can also have a significant impact on the results. In fact, assigning greater weights to samples that are difficult to segment is not always a good strategy. Rather than focusing on the cases of samples that are difficult to segment, the proposed method is primarily concerned with learning from minority samples, based on the number of points present during the training process. However, the weighted cross-entropy function actually focuses more on the accuracy of individual classes, rather than the errors within specific classes. This means that if greater weights are allocated to minority samples, the number of errors in these classes could also be larger. Therefore, the proposed loss function is designed to balance the performance of minority classes with the overall performance across classes. The proposed weighted cross-entropy is as follows:

$$L = -\sum_{c=1}^{M} \frac{\prod_{i=1,\, i \neq c}^{M} \sqrt{N_i}}{\sum_{i=1}^{M} \sqrt{N_i}} y_c \log\left(p_c\right) \tag{7.13}$$

where N_c is the total point number of cth class present in the training process and M denotes the number of classes, while y_c and p_c denote the true label vector and predicted label vector of cth class respectively.

7.2 EXPERIMENTS

7.2.1 Implementation and Dataset

Due to the large number of points and the large scale, to accelerate computation using batches, the entire scene is divided into 10 m × 10 m blocks (Block) during training, each block being randomly sampled to 20,000 points, which may include duplicate points, while using a neighborhood range with a radius of 0.8 m for normal vector calculation. The input point features include global coordinates, normalized coordinates within the block, 0-1 normalized coordinates, normals, and intensity. To enrich the training dataset, there is a 5 m overlap between every two adjacent blocks, so the number of predicted points is usually greater than the number of initially scanned points in that area. The method proposed in this chapter uses 0.05 m as the resolution for rule-based sampling before blocking, followed by five downsampling layers with sampling ratios of 0.25, 0.25, 0.25, 0.25, 0.25, 0.5, and 0.5, respectively. The network is implemented on the PyTorch platform. During training, the Adam optimizer is used to update the model, with momentum and initial learning rate set at 0.9 and 0.001, respectively, and the decay rate set at 0.0001, with the learning rate halving every 16 epochs. The model was trained on an NVIDIA GTX 1080Ti GPU for 100 iterations with a batch size of 28, and the model with the best mIoU was selected for testing. During testing, the trained model can predict 10^6 points in a single forward pass using a single GPU (11G of memory).

WHU-Urban3D Shanghai MLS dataset (Han et al. 2024) and Toronto-3D dataset (Tan et al. 2020) are used for the experiment. The Shanghai MLS dataset encompasses 40 scenes, amassing more than 300 million points in total, with 30 scenes designated for training and ten for evaluation. Various categories within the dataset have been explicitly annotated, and Figure 7.3 illustrates the distribution of select categories.

- Ground: vehicular lanes (excluding road signs), non-vehicular lanes (non-lane ground areas not belonging to lanes), and road markings;
- Construction: buildings (encompassing building facades and other miscellaneous items within buildings) and fence (including road isolation structures and fences, etc.);

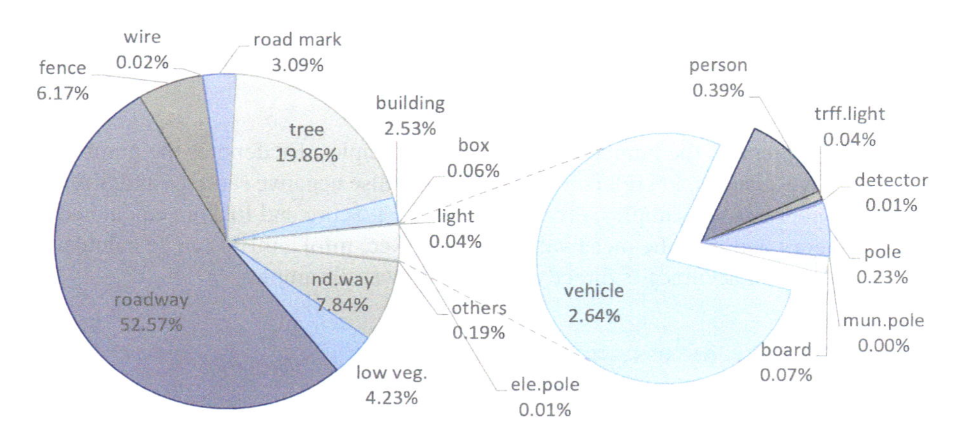

FIGURE 7.3 The proportion of different categories in Shanghai MLS dataset. (Han et al. 2024).

- Vegetation: trees and low vegetation (short plants, including grass, shrubs, and other low trees);
- Dynamic Objects: pedestrians and vehicles. Objects like bicycles and motorbikes are also included in the pedestrian category.
- Poles: light, tel. pole (telegraph pole), mun. pole (municipal pole), trff. light (traffic light), detector, board (usually attached to the light).

The Toronto-3D dataset is divided into four parts, spanning nearly one kilometer of roadway and comprising around 80 million data points. This dataset features eight distinct classifications, including road, road markings (rd. mrk.), natural objects, building, utility line (util. line), pole, vehicle and fence. Each point is characterized by its position, color, intensity, GPS time and scan angle rank. In the experiment, the second scene serves as the test set while others for training as in (Tan et al. 2020). Notably, the experiment leverages only the spatial coordinates (x, y, z) and brightness levels as inputs.

Precision (Pre), recall (Rec), $F1$-Score, IoU and overall accuracy (OA) are used in this chapter for evaluation and are defined as follows.

$$\text{Pre} = \frac{\text{TP}}{\text{TP} + \text{FP}} \tag{7.14}$$

$$\text{Rec} = \frac{\text{TP}}{\text{TP} + \text{FN}} \tag{7.15}$$

$$F1 - \text{Score} = 2 \times \frac{\text{Pre} \times \text{Rec}}{\text{Pre} + \text{Rec}} \tag{7.16}$$

$$\text{IoU} = \frac{\text{T} \cap \text{P}}{\text{T} \cup \text{P}} = \frac{\text{TP}}{\text{TP} + \text{FP} + \text{FN}} \tag{7.17}$$

$$\text{OA} = \frac{\sum \text{TP}}{N} \tag{7.18}$$

where TP represents the number of true positive samples, FP denotes the number of false positive samples, FN refers to the number of false negative samples and N refers to the number of total samples. Precision, recall, $F1$-Score and IoU are calculated in each category and then the mean value (mPre, mRec, mIoU, mF1) can be calculated while the overall accuracy is directly calculated over all samples.

7.2.2 Semantic Segmentation Results

Figure 7.4 illustrates the prediction outcomes for different testing scenes. Specifically, Figure 7.4(a) highlights a scene with intricate structures unique to the test data, a rarity within the Shanghai MLS dataset, yet the prediction performance is commendable. Figure 7.5 showcases some categories from a close-up view. According to Figure 7.6(a), within the test scenes, the categories of detector, traffic light, low vegetation, municipal pole, telegraph pole, and box are prone to being incorrectly identified as traffic light, board, tree, pedestrian, traffic light, and fence, respectively. Conversely, the points wrongly classified as box, telegraph pole, municipal pole, low vegetation, board, traffic light, detector, and wire are probably actually fence, tree, non-drive way, tree, light, telegraph pole, wire, and building.

These classification errors can be attributed to three primary factors: (1) The presence of semantically ambiguous classes, where a point could belong to more than one

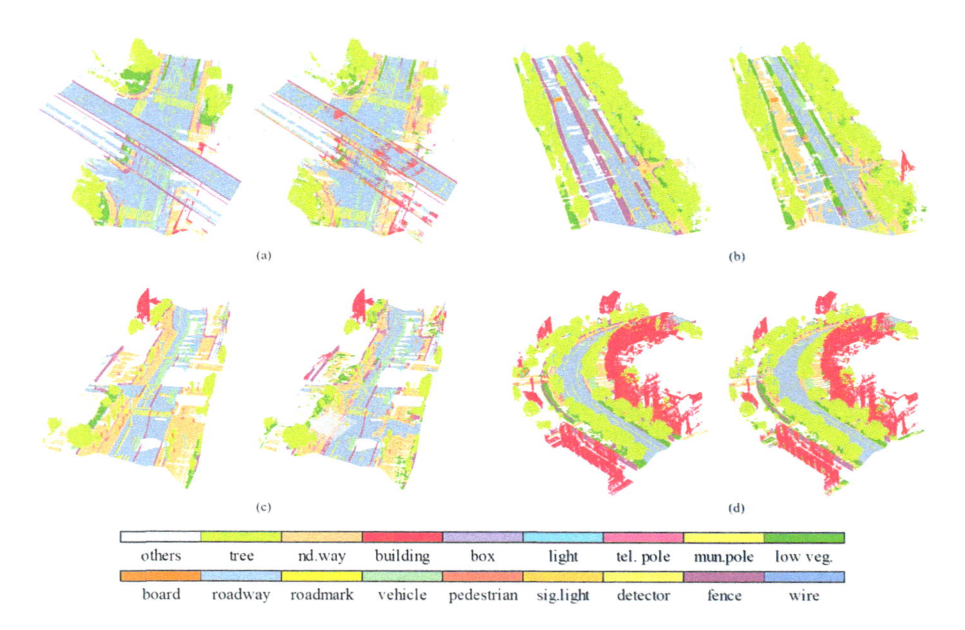

FIGURE 7.4 Prediction results on four scenes of Shanghai MLS dataset. Manually labeled results are in the left and predicted ones in the right. (Han, Dong, and Yang 2021).

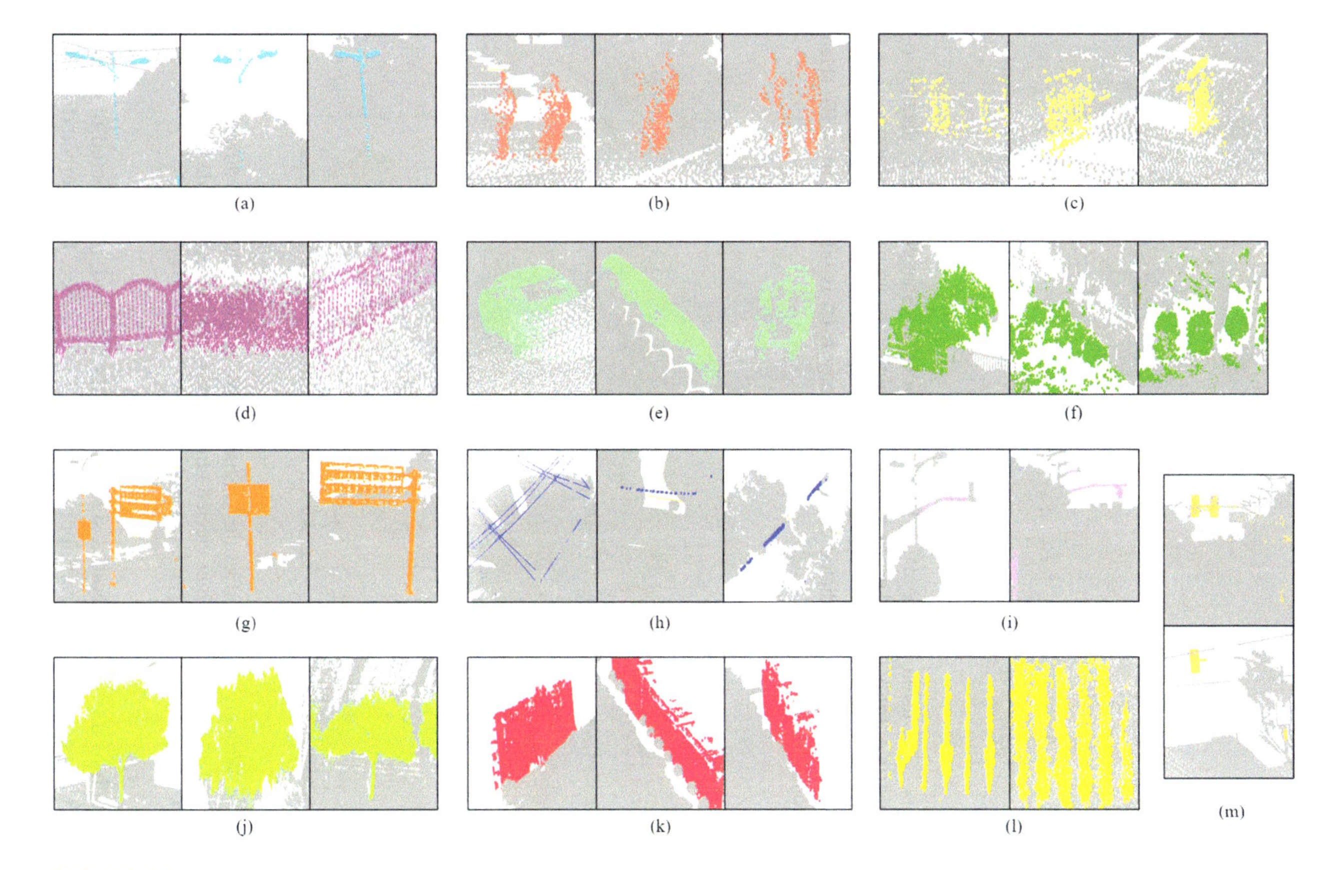

FIGURE 7.5 Predicted results of some specific categories of Shanghai MLS dataset: (a) light; (b) pedestrian; (c) box; (d) fence; (e) vehicle; (f) low vegetation; (g) board; (h) wire; (i) detector; (j) tree; (k) building; (l) road mark; (m) traffic light. (Han, Dong, and Yang 2021).

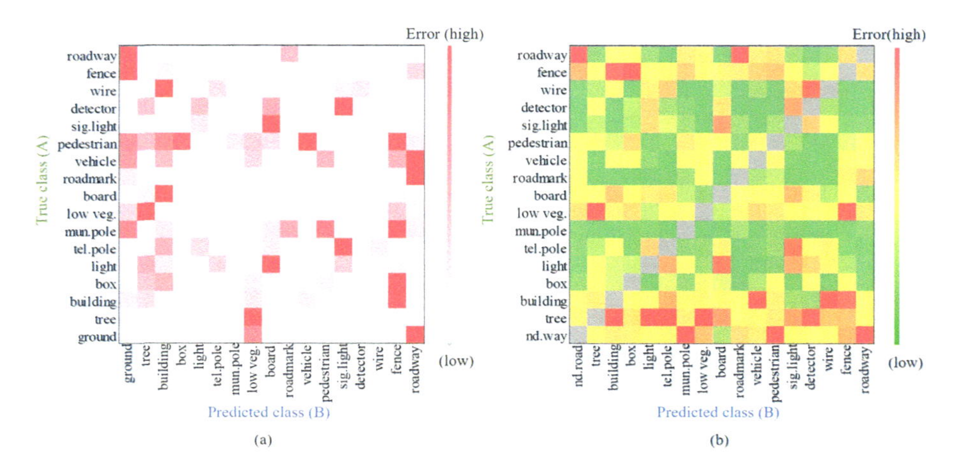

FIGURE 7.6 Confusion matrix of different categories in Shanghai MLS dataset. (a) Confusion matrix of truth: the darker the red color, the higher the probability that true class A will be wrongly predicted as class B. (b) Confusion matrix of prediction: the redder the color, the higher the probability of predicted class B is actually A (A is different from B). (Han, Dong, and Yang 2021).

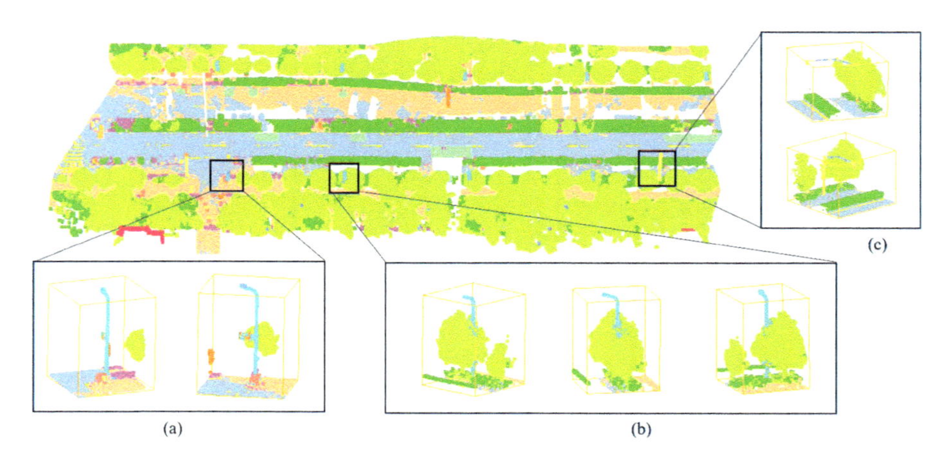

FIGURE 7.7 Wrongly segmented results of Shanghai MLS dataset shown in different views. (a) Board attached to light. (b) Light elected among trees. (c) Detector divided by different blocks. (Han, Dong, and Yang 2021).

category, such as a pole with both a sign and a light attached, as depicted in Figure 7.7(a). (2) The occurrence of overlapping objects, for instance, lights erected among trees, as illustrated in Figure 7.7(b). (3) The variation in predicted classes at specific localities is possibly due to divergent block predictions, as shown in Figure 7.7(c).

This chapter employs hierarchical IoU and $F1$-Score evaluation metrics on the Shanghai MLS dataset to account for varying levels of classification. Remarkably, a single trained model suffices for this multi-tiered evaluation, eliminating the need for distinct models tailored to each category. Predicted labels can be effortlessly aligned with their respective hierarchical levels by label mapping according to specific demands. Table 7.1 delineates the segmentation outcomes across all scenes for various categories. Aggregate metrics including mean IoU, $F1$-Score, precision, recall, and overall accuracy are computed across all points, transcending individual scene distinctions. This encompasses 17 fine categories (such as tree, non-drive way, building, box, light, telegraph pole, municipal pole, low vegetation, board, roadway, road mark, vehicle, pedestrian, traffic light, detector, fence, and wire) and six coarser categories (dynamic object, pole facility, vegetation, construction,

TABLE 7.1
Semantic segmentation Results of Shanghai MLS Dataset

Class		mIoU (%)	F1-Score (%)	Precision (%)	Recall (%)
person		60.8	75.6	77.9	73.4
car		79.1	88.3	90.5	86.3
dynamic		**77.5**	**87.3**	**89.2**	**85.6**
light		71.8	83.6	79.9	87.7
sig.light		31.0	47.3	46.0	48.6
detector		31.3	47.7	40.6	57.8
ele.pole		49.9	66.6	64.8	68.4
mul.pole		26.5	41.8	35.6	50.8
board		20.2	33.7	31.1	36.7
pole facilities		**72.0**	**83.7**	**72.0**	**100.0**
tree		84.5	91.6	88.8	94.5
low veg.		34.1	50.8	52.9	48.9
vegetation		**86.5**	**92.8**	**90.6**	**95.0**
fence		57.9	73.3	90.9	61.5
building		77.1	87.1	85.2	88.9
construction		**72.5**	**84.1**	**91.5**	**77.7**
driveway		83.6	91.0	94.8	87.6
nd. way		58.4	73.7	58.4	100.0
roadmark		38.1	55.2	52.1	58.6
ground		**87.8**	**93.5**	**91.7**	**95.4**
box		45.4	62.5	65.0	60.1
wire		47.2	64.1	63.9	64.3
other utilities		**46.8**	**63.8**	**65.6**	**62.0**
mean	**coarse**	**73.9**	**84.2**	**83.4**	**85.9**
	fine	**52.7**	**66.7**	**65.8**	**69.1**

ground, and other utilities, as highlighted in Table 7.1 in bold), all of which span the entire scene. The method demonstrates robust performance in fine classifications for larger entities like driveways (IoU: 83.6, F1-Score: 91.0), buildings (IoU: 77.1, F1-Score:87.1), trees (IoU: 84.5, F1-Score: 91.6), and dynamic objects such as pedestrians (IoU: 60.8, F1-score: 75.6) and vehicles (IoU: 79.1, F1-Score: 88.3), while also delivering commendable results in coarser categories, achieving a mean IoU of 73.9 and mean F1-Score of 84.2. It is important to note that both fine and coarse categories encompass the entire scene using the identical set of points. Despite covering identical areas, coarse categories typically yield higher performance due to the propensity for subcategories to be mistakenly interchanged, a phenomenon also illustrated by Figure 7.6. This can be attributed primarily to two factors: (1) entities within the same coarse category share similar attributes (for instance, roadway and non-drive way); (2) entities are intricate constructions with ambiguous definitions (like light and board). Table 7.2 presents the efficacy of various methods across 17 categories, with the proposed method achieving superior outcomes and surpassing other techniques in numerous categories, including non-drive way, building, box, light, telegraph pole, municipal pole, board, roadway, road mark, vehicle, pedestrian, detector, and wire. Table 7.3 details the parameter count and inference times across different layers, with calculations based on processing a million points per forward pass. The network, with a total of 3.7 million parameters, is capable of processing a million points within two seconds, showcasing its efficiency and lightweight design.

In the case of Toronto-3D, Table 7.4 presents the testing performance, showcasing the superior performance of our proposed method in terms of both mean IoU and overall accuracy. Specifically, our method outperforms alternative models, particularly in segments like road mark, vehicle, road, and natural objects. The comparative results of other methods on Toronto-3D are sourced from (Tan et al. 2020).

7.3 ANALYSIS AND DISCUSSION

7.3.1 Sampling Strategy Analysis

This section evaluates various elementary sampling strategies utilized in MLS point cloud data, including: (1) Random Sampling (RS); (2) Farthest Point Sampling (FPS); (3) A combination of RS and FPS; (4) A blend of Uniform Sampling (US), RS, and FPS, with each method being indifferent to the class of the samples. As depicted in Figure 7.8, despite initial uniform sampling of the input point set at a 0.05 resolution, the random sampling approach exhibits a preference for densely populated areas (such as the ground and trees), while sparsely-populated areas (like lights) are often overlooked. Consequently, objects in these less dense areas are difficult to detect by the 5th layer, as their points are almost entirely excluded from the sample. In contrast, merging farthest point sampling with random sampling results in a more balanced selection of points, thus preserving the integrity of these objects to a certain

TABLE 7.2
The Achieved IoU (%) by Different Methods on Shanghai MLS Dataset

Methods	Mean	Tree	nd.Way	Building	Box	Light	tel.Pole	mun.Pole	Low veg	Board
		Roadway	rd.mrk.	Vehicle	Pedestrian	trff.Light	Detector	Fence	Wire	
PointNet++	41.1	83.3	42.0	72.7	6.6	59.1	30.8	7.8	33.1	13.9
		80.0	29.5	76.7	38.9	25.0	11.0	56.3	32.7	
PointConv	46.4	85.6	48.9	73.5	28.2	59.7	35.7	20.0	32.4	16.0
		82.0	30.6	76.2	53.8	28.7	27.6	52.6	36.5	
RandLA	47.1	86.8	55.6	76.3	32.4	61.2	43.1	15.2	37.4	18.1
		80.6	29.3	74.9	50.7	21.6	22.2	55.8	39.2	
Proposed	52.8	84.5	58.4	77.1	45.4	71.8	49.9	26.5	34.1	20.2
		83.6	38.1	79.1	60.8	31.0	31.3	57.9	47.2	

TABLE 7.3
Parameters and Mean Time of Inference of one Million Points with One GPU

Layer	Parameters	Mean time(s)
sa1	6K	0.299
sa2	27K	0.278
sa3	103K	0.365
sa4	402K	0.113
sa5	1592K	0.068
fp5	1051K	0.052
fp4	263K	0.008
fp3	165K	0.014
fp2	115K	0.045
fp1	50K	0.181
Fc	19K	0.001
Total	3.7M	1.424

degree. According to Table 7.5, solely employing FPS for spatial downsampling with an input of 20,000 points requires over five minutes, whereas the hybrid approach significantly reduces this time to merely one second, making it feasibly efficient for experimental purposes. Considering both efficiency and efficacy, the hybrid method is selected for deployment.

7.3.2 Feature Aggregation Module Analysis

To validate the efficacy of the aggregation strategy, this section compares the proposed approach with various operations and features, as outlined in Table 7.5. Three methods of grouping local features are evaluated: (1) Utilizing solely relative features; (2) Employing solely absolute features; and (3) Integrating both relative and absolute features. Additionally, the section examines different inputs of C for the spatial activation function, focusing on the comparison of using relative coordinates versus other methods. Results indicate that utilizing relative coordinates as input yields superior effectiveness compared to other approaches, whereas employing only relative features or combining them with relative coordinates demonstrates weaker performance, akin to conventional practices in attentional neural networks. Furthermore, the study compares different aggregation methods, revealing that employing max pooling as the aggregating operation confers advantages over average pooling and summation operations.

TABLE 7.4
Comparison of Semantic Segmentation Results of Several Methods on Toronto-3D

Methods	IoU (%)									OA (%)
	Mean	Road	Rd mrk.	Natural	Building	Util. line	Pole	Car	Fence	
PointNet++	56.6	91.4	7.6	89.8	74.0	68.6	59.5	54.0	7.5	91.2
PointNet++ (MSG)	53.1	90.7	0.0	86.7	75.8	56.2	60.9	44.5	10.2	90.6
DGCNN	49.6	90.6	0.4	81.3	64.0	47.1	56.9	49.3	7.3	89.0
KPFCNN	60.3	90.2	0.0	86.8	**86.8**	**81.1**	**73.1**	42.9	**21.6**	91.7
MS-PCNN	58.0	91.2	3.5	90.5	77.3	62.3	68.5	53.6	17.1	91.5
TGNet	58.3	91.4	10.6	91.0	76.9	68.3	66.3	54.1	8.2	91.6
MS-TGNet	61.0	90.9	18.8	92.2	80.6	69.4	71.2	51.1	13.6	91.7
Proposed	**70.8**	**92.2**	**53.8**	**92.8**	86.0	72.2	72.5	**75.7**	21.2	**93.6**

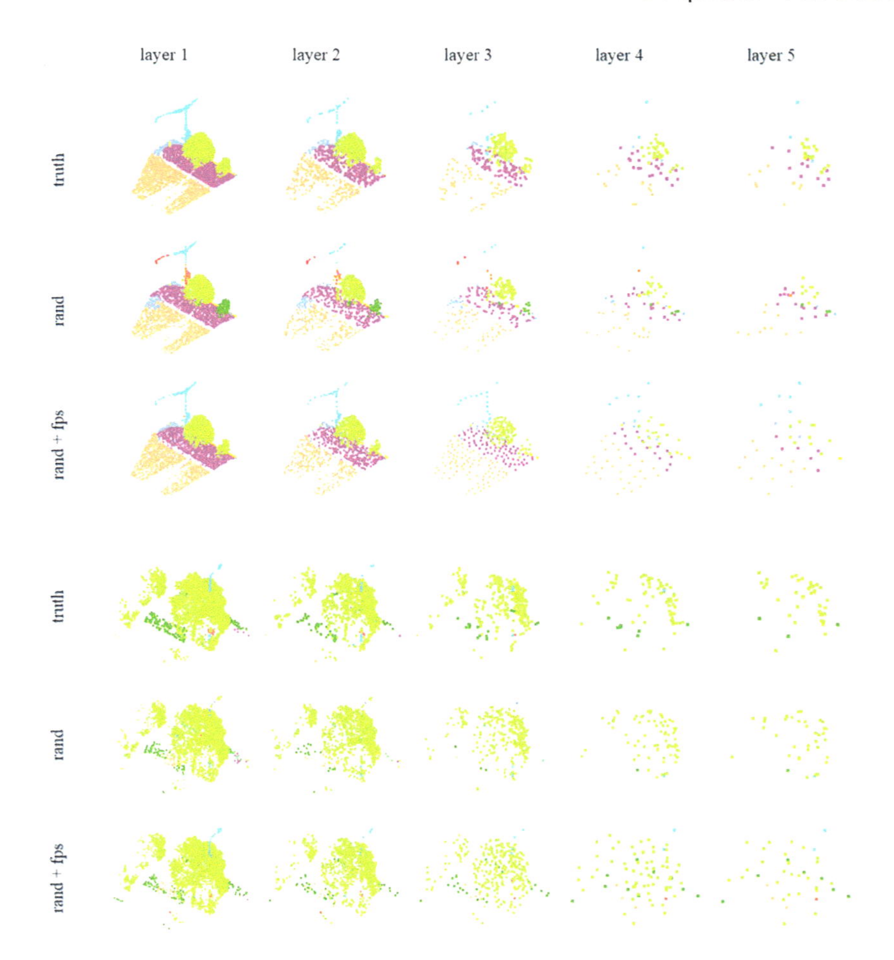

FIGURE 7.8 Comparison between different sampling strategies. The number of sampled points of 1th, 2th, 3th, 4th and 5th are 5000, 1250, 312, 78, 39 respectively. The color indicates the prediction result of the corresponding sampling method. (Han, Dong, and Yang 2021).

TABLE 7.5
Time of Different Sampling Strategies

Layer		Layer 1	Layer 2	Layer 3	Layer 4	Layer 5	
Number of points		**20000**	**5000**	**1250**	**312**	**78**	**Total**
	RS	0.0794	0.0080	0.0026	0.0017	0.0009	0.0926
Time(s)	FPS	286.7074	15.3844	0.9264	0.0696	0.0173	303.1051
	Combined	0.0758	0.0082	0.9186	0.0684	0.0181	1.0891

All the methods regularly sample the point set first and use batch calculation to compute 10^6 points for acceleration.

7.3.3 Loss Function Analysis

To address class imbalance, this section explores various methods for defining class weights. These weights are determined based on the number of points in each class during training, including N_c, $\sqrt{N_c}$ and $\log N_c$. The reference cross-entropy is then defined as:

$$L_{\text{ce}} = -\sum_{c=1}^{M}\left[y_c \log\left(p_c\right)\right] \tag{7.19}$$

The weighted cross entropy is defined as:

$$L_{\text{ce}-\{\text{se,lo,sq}\}} = -\sum_{c=1}^{M} w_{c,\{\text{se,lo,sq}\}} y_c \log\left(p_c\right) \tag{7.20}$$

$$w_{\text{se},c} = \frac{1}{N_c \sum_{i=1}^{M} \frac{1}{N_i}} \tag{7.21}$$

$$w_{\text{sq},c} = \frac{1}{\sqrt{N_c} \sum_{i=1}^{M} \frac{1}{\sqrt{N_i}}} \tag{7.22}$$

$$w_{\text{lo},c} = \frac{1}{\log N_c \sum_{i=1}^{M} \frac{1}{\log N_i}} \tag{7.23}$$

N_c represents the count of class c instances observed during training. Illustrated in Figure 7.9, $L_{\text{ce}-\text{lo}}$ exhibits the swiftest convergence within the training dataset, trailed by $L_{\text{ce}-\text{sq}}$, with the converged loss smaller than other functions. Conversely, in the testing dataset as shown in Figures 7.10 and 7.11, although the mIoU of $L_{\text{ce}-\text{sq}}$ achieves optimal performance despite a slower decline compared to other methods, overall accuracy in testing data remains consistent across loss functions L_{ce}, $L_{\text{ce}-\text{sq}}$ and $L_{\text{ce}-\text{lo}}$. Notably, directly assigning weights as the inverse of point numbers results in the poorest outcomes in both training and testing datasets. This can be attributed to the inherent difficulty of learning many minority classes, characterized by low occupancy and complex structures. Assigning higher weights to minority classes may exacerbate overfitting with poor-quality samples, thus hindering network-wide learning. Focal loss (Lin et al. 2017) primarily addresses the imbalance between positive and negative samples in single-stage object detection. This loss function alleviates the influence of numerous straightforward negative samples during training, akin to a form of challenging sample-mining. The original focal loss is defined as described in (Lin et al. 2017).

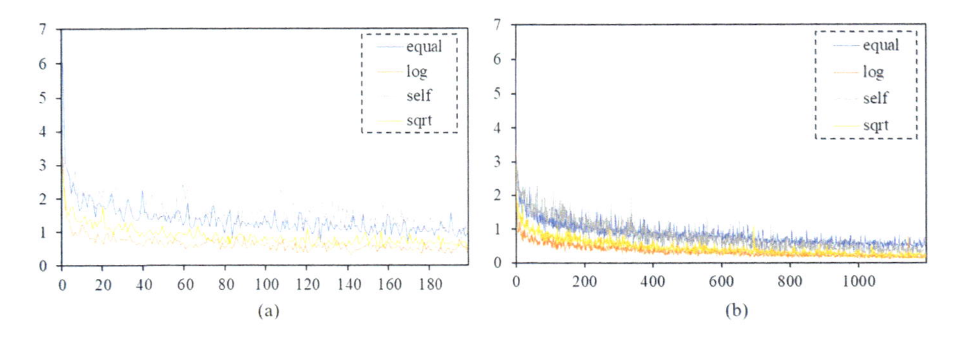

FIGURE 7.9 Losses on training dataset with different loss functions. (a) The first iteration. (b) The first 80 iterations. Equal, log, self, and sqrt represent L_{ce}, L_{ce-lo}, L_{ce-se} and L_{ce-sq}, respectively.

$$\mathrm{L}_{fl} = \begin{cases} -\alpha(1-y')^{\gamma}\log y' & y=1 \\ -(1-\alpha)y'^{\gamma}\log(1-y') & y=0 \end{cases} \tag{7.24}$$

Here, α represents a factor for balancing positive and negative samples. γ serves as a tunable focusing parameter, with $\gamma \geq 0$, and $(1 - y')$ acts as a modulating factor for cross-entropy loss. Similarly, the loss for multi-class segmentation can be defined as follows:

$$L_{fl-seg} = -\sum_{c=1}^{M}\left[\alpha(1-p_c)^{\gamma} y_c \log p_c\right] \tag{7.25}$$

Considering the extensive range of categories compared to (Lin et al. 2017), which solely deals with positive and negative samples (treated as two categories), determining a suitable α through experimentation becomes challenging. Consequently, α is directly set as w, akin to weighted cross-entropy, while γ is fixed at 2, mirroring (Lin et al. 2017). Despite the dynamic adaptation of weights by focal loss based on calculated losses in each iteration, this training approach falls short of optimizing overall model performance, as depicted in Figure 7.10.

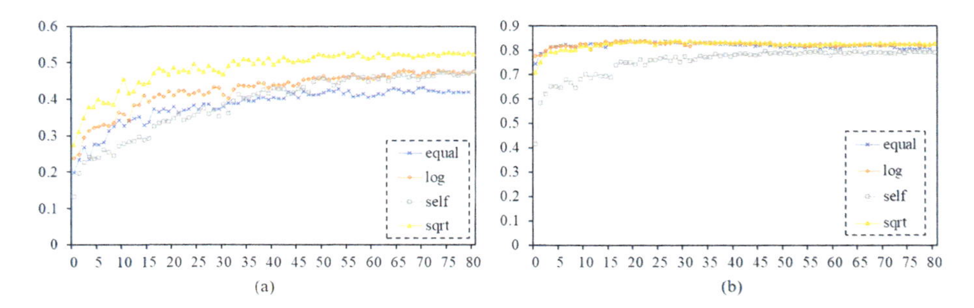

FIGURE 7.10 Performance on validation dataset with different loss functions. (a) Mean IoU. (b) Overall accuracy.

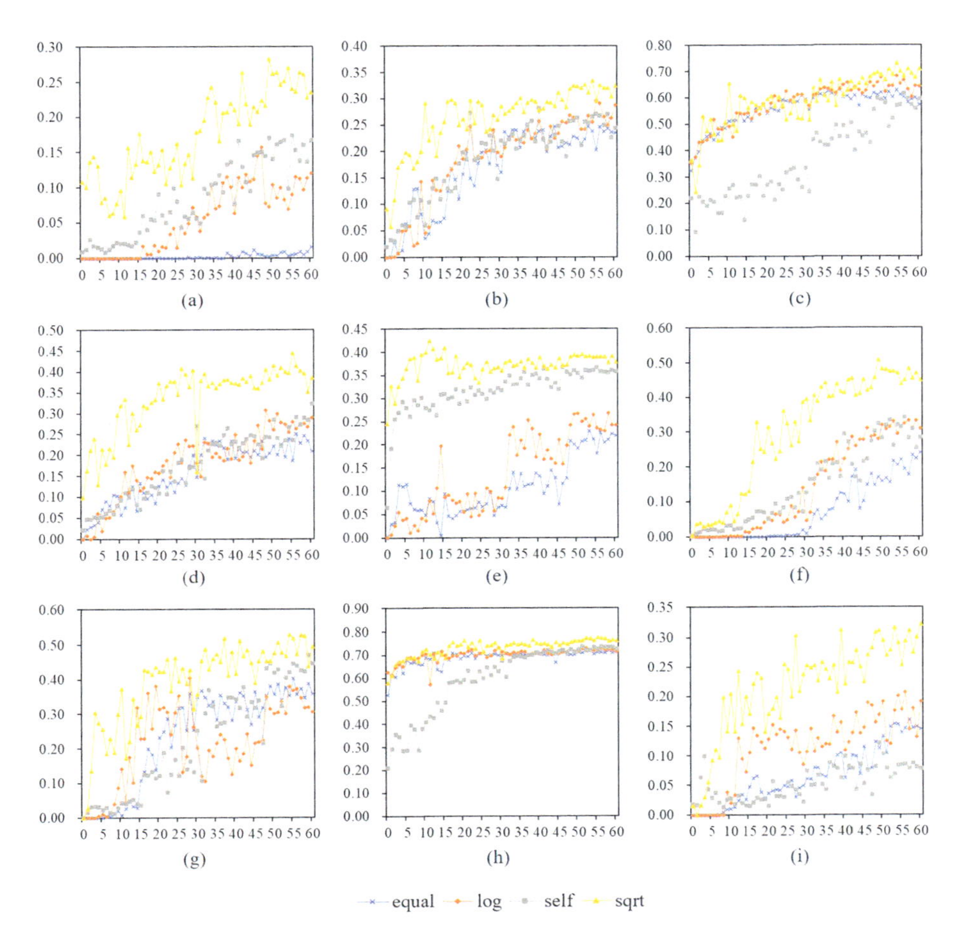

FIGURE 7.11 Iterative IoU of some categories with different loss functions: (a) municipal pole; (b) traffic light; (c) light; (d) wire; (e) road mark; (f) box; (g) telegraph pole; (h) building; (i) detector.

7.4 CONCLUSION

This chapter introduces a deep learning based method for semantic segmentation of large-scale MLS point clouds, focusing on spatial sampling, local feature aggregation, and loss function. Initially, an end-to-end deep neural network is proposed for per-point classification of outdoor large-scale point clouds, efficiently segmenting the entire road scene into nearly 20 categories without pre-processing or post-processing. Subsequently, an efficient spatial downsampling strategy is implemented, followed by the introduction of a point-wise feature aggregation module that is permutation invariant and translation invariant, aiming to encode local geometry information effectively. Additionally, a designed loss function addresses severe category imbalance, significantly enhancing overall performance. Experimental results demonstrate the effectiveness and efficiency of the proposed method in segmenting challenging large-scale road scenes, achieving state-of-the-art mean IoU of 70.8% and 73.9% in the Toronto-3D and Shanghai MLS datasets, respectively. Notably, improvements in IoU for specific categories such as boxes, lights, road markings, and pedestrians are observed, with notable enhancements in the Shanghai MLS dataset.

However, a limitation arises from the division of scenes into blocks before feeding points into the network, resulting in predictions solely based on features within one block, thus lacking interblock relationships. Moreover, the segmentation of MLS datasets remains challenging, particularly concerning small object detection and handling complex structures attached by different objects like lights and boards, due to ambiguous category definitions. Future research avenues could explore label prediction without blocks or with larger ones, or investigate learning related features between blocks to potentially enhance predictions, possibly integrating image data with point clouds for improved small object segmentation.

REFERENCES

Han, Xu, Zhen Dong, and Bisheng Yang. 2021. "A Point-Based Deep Learning Network for Semantic Segmentation of MLS Point Clouds." *ISPRS Journal of Photogrammetry and Remote Sensing* 175: 199–214. doi:10.1016/j.isprsjprs.2021.03.001

Han, Xu, Chong Liu, Yuzhou Zhou, Kai Tan, Zhen Dong, and Bisheng Yang. 2024. "WHU-Urban3D: An Urban Scene LiDAR Point Cloud Dataset for Semantic Instance Segmentation." *ISPRS Journal of Photogrammetry and Remote Sensing* 209: 500–513. doi:10.1016/j.isprsjprs.2024.02.007

Hermosilla, Pedro, Tobias Ritschel, Pere-Pau Vázquez, Àlvar Vinacua, and Timo Ropinski. 2018. "Monte Carlo Convolution for Learning on Non-Uniformly Sampled Point Clouds." *ACM Transactions on Graphics (TOG)* 37 (6): 1–12. doi:10.1145/3272127.3275110

Lin, Tsung-Yi, Priya Goyal, Ross Girshick, Kaiming He, and Piotr Dollár. 2017. "Focal Loss for Dense Object Detection." In *Proceedings of the IEEE International Conference on Computer Vision*, 2980–88. doi:10.1109/iccv.2017.324

Qi, Charles R, Hao Su, Kaichun Mo, and Leonidas J Guibas. 2017. "Pointnet: Deep Learning on Point Sets for 3d Classification and Segmentation." In *Proceedings of the IEEE Conference on Computer Vision and Pattern Recognition*, 652–60. doi:10.1109/cvpr.2017.16

Tan, Weikai, Nannan Qin, Lingfei Ma, Ying Li, Jing Du, Guorong Cai, Ke Yang, and Jonathan Li. 2020. "Toronto-3D: A Large-Scale Mobile LiDAR Dataset for Semantic Segmentation of Urban Roadways." In *Proceedings of the IEEE/CVF Conference on Computer Vision and Pattern Recognition Workshops*, 202–3. doi:10.1109/cvprw50498.2020.00109

Wu, Wenxuan, Zhongang Qi, and Li Fuxin. 2019. "Pointconv: Deep Convolutional Networks on 3d Point Clouds." In *Proceedings of the IEEE/CVF Conference on Computer Vision and Pattern Recognition*, 9621–30. doi:10.1109/cvpr.2019.00985

Xu, Yifan, Tianqi Fan, Mingye Xu, Long Zeng, and Yu Qiao. 2018. "Spidercnn: Deep Learning on Point Sets with Parameterized Convolutional Filters." In *Proceedings of the European Conference on Computer Vision (ECCV)*, 87–102. doi:10.1007/978-3-030-01237-3_6

8 Point Cloud Instance Segmentation

8.1 METHODOLOGY

8.1.1 Overview

The proposed point cloud instance segmentation method integrates street-view imagery and mobile laser scanning (MLS) point clouds as input. It leverages both semantic and 3D geometric information and outputs component-level street furniture locations with detailed semantic labels and point-level instance masks. Generally, as shown in Figure 8.1, this method consists of three major steps. Firstly, based on the pre-established point-image alignment and the detected street furniture objects, point cloud frustums-of-interest (FoI) are cropped. Then, a multi-task neural network is used to predict the instance bounding box and the point mask for each FoI. In the final phase, the predicted bounding boxes are used to link identical instances across frustums, leading to the final object-centric instance masks via fusing the instance mask prediction results of those associated frustums.

8.1.2 Frustum-of-interest Cropping

The alignment of street-view imagery and point clouds from the MLS system is pivotal for integrating 2D and 3D information. This alignment is established using camera coordinates and Euler angles of picturing moments and pre-calibrated parameters. This alignment maps a 3D point (x, y, z) from the local projection coordinate system (x_p, y_p, z_p) to the camera-centric coordinate system (x_c, y_c, z_c), and then to the 2D panoramic pixel coordinate system via spherical projection. Therefore, each 3D point can be mapped to an associated pixel on the image plane.

Off-the-shelf image object detection models are adopted to generate 2D proposals and lift them to 3D point cloud spaces for frustum cropping. The state-of-the-art model on benchmark Objects 365 (Shao et al. 2019) is used for the experiments introduced in this chapter due to its good performance in detecting a wide range of street furniture objects from panoramic images. Further, to expand the range of traffic sign detection beyond the basic types in Objects 365, we refine the categories using TT100k dataset by fine-tuning a lighter detection model with it (Zhu et al. 2016). In total, the surveyed street furniture in this setup consists of common transportation infrastructures including traffic signs, traffic lights, street lamps and temporary cone barriers, and extends to communal assets like fire hydrants, trash bins, potted plants, and roadside benches. Moreover, traffic signs are further detailed into dozens of categories including text signs, stop signs, and various warning or speed limit signs.

DOI: 10.1201/9781003486060-11

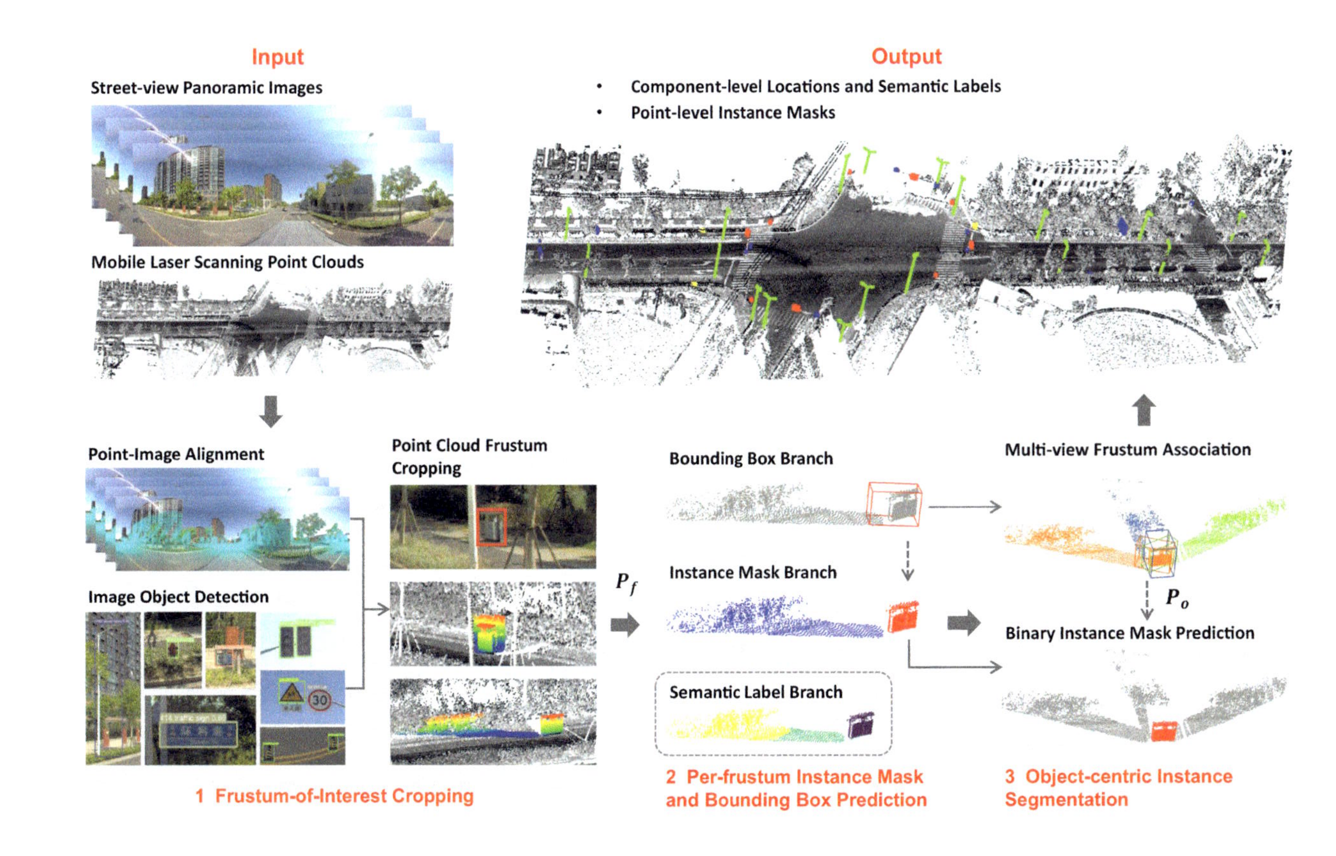

FIGURE 8.1 The pipeline of the proposed method (Zhou et al. 2022).

Based on the previously determined alignment, each detected 2D bounding box is lifted to a 3D frustum-of-interest in the point cloud space and the points within the frustum constitute a frustum point cloud $\boldsymbol{P}_f \in \mathbb{R}^{N \times (3+c)}$, where N is the point number and c is the number of additional feature channels (e.g., intensity, RGB). Then, a coordinates rotation of $\boldsymbol{P}_f$ along the y-axis, making the z-axis pointing to the frustum center, is performed for each frustum to standardize the position and minimize the variability of the point clouds within each frustum (Qi et al. 2018). This reorientation transforms the points into the dedicated frustum coordinate system (x_f, y_f, z_f).

8.1.3 Per-frustum Instance Mask and Bounding Box Prediction

In this phase, the objective is to predict instance masks measuring the probability of being street furniture points for each frustum point cloud $\boldsymbol{P}_f$. It is assumed that a single frustum-of-interest (FoI) encapsulates only one street furniture instance (i.e., the detected object) and other points are considered as non-relevant points. A baseline idea is to directly predict the instance mask from point features, but such a method is prone to errors especially false positives, because it pays overwhelming focus on the point-level local features and lacks the contextual awareness of the instance.

To mitigate this issue, we concurrently estimate a 3D bounding box of the road furniture and use it as an input of the instance mask prediction, to enhance the contextual awareness of the detected instance. Another merit of the bounding box prediction is that it supports the subsequent procedure of FoI fusion by providing concrete information about the location and size of the object. Meanwhile, the point-level semantic prediction branch further promotes point feature learning. The above tasks are designed to be mutually reinforcing in this network and the experiments have proved this hypothesis. Specifically, we adopt a neural network containing three branches to process each frustum point cloud $\boldsymbol{P}_f$: (a) bounding box branch; (b) instance mask branch; (c) semantic label branch. And they share a point cloud processing backbone, as shown in Figure 8.2.

Bounding Box Branch. This branch takes the global feature $\boldsymbol{F}_g$ as input and outputs min-max coordinates $\boldsymbol{B} = [x_{\min}, y_{\min}, z_{\min}, x_{\max}, y_{\max}, z_{\max}]$ of the estimated bounding box. $\boldsymbol{F}_g$ concatenates the point cloud global feature $\boldsymbol{F}_{g-pc}$ and the one-hot semantic vector $\boldsymbol{F}_{\mathbf{img}}$ from image object detection.

We use $\mathcal{L}_{box}$ in Eq (8.1) to supervise this branch.

$$\mathcal{L}_{box} = \mathcal{L}_{dis} + \mathcal{L}_{siou} \tag{8.1}$$

In Eq. (8.1), $\mathcal{L}_{dis}$ measures the spatial coordinate differences and $\mathcal{L}_{siou}$ encourages the predicted box to include more valid instance points. $\mathcal{L}_{dis}$ in Eq. (8.2) measures the Euclidean distance between the predicted bounding box $\boldsymbol{B}$ and the ground truth bounding box $\bar{\boldsymbol{B}}$.

$$\mathcal{L}_{dis} = -\frac{1}{6}\Sigma\left(\boldsymbol{B} - \bar{\boldsymbol{B}}\right)^2 \tag{8.2}$$

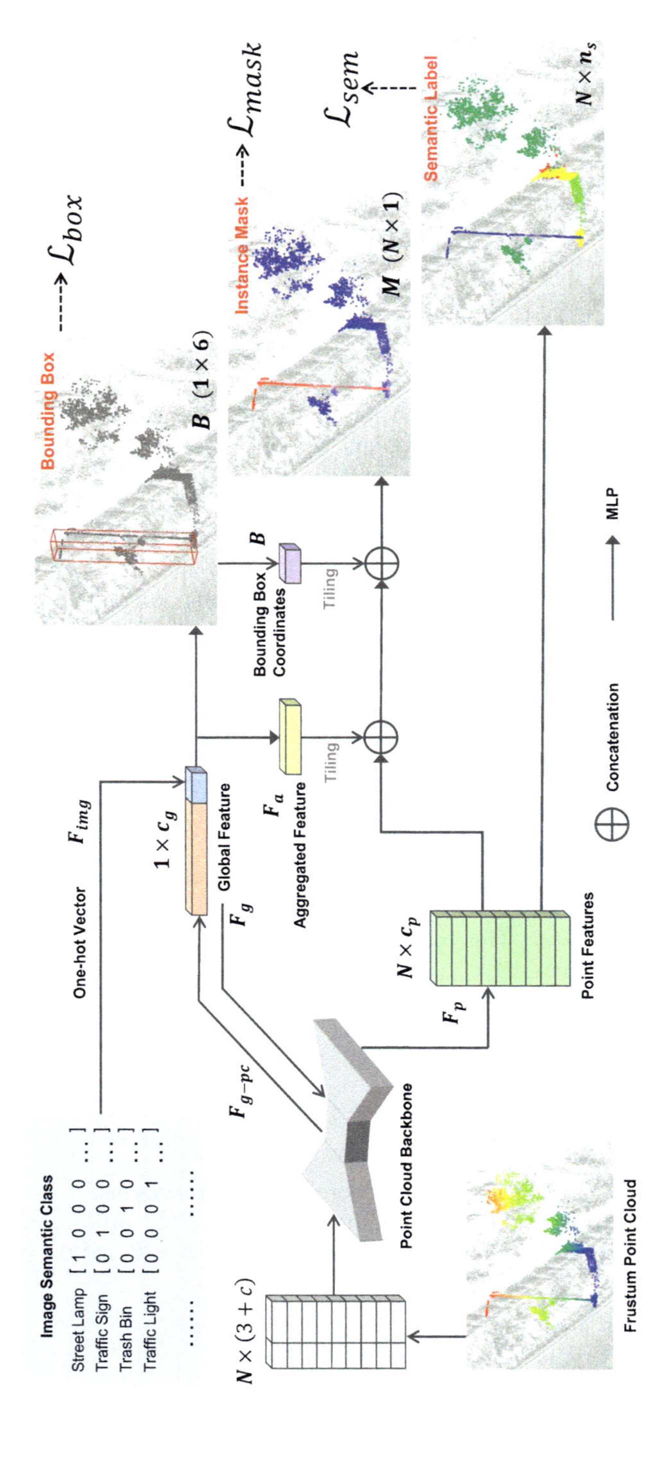

FIGURE 8.2 The framework of the multi-task neural network (Zhou et al. 2022).

And $\mathcal{L}_{siou}$ in Eq. (8.3) is a soft intersection-over-union (sIoU) loss introduced by (Yang et al. 2019).

$$\mathcal{L}_{siou} = \frac{-\sum_{n=1}^{N}\left(q_n * \bar{q}_n\right)}{\sum_{n=1}^{N} q_n + \sum_{n=1}^{N} \bar{q}_n - \sum_{n=1}^{N}\left(q_n * \bar{q}_n\right)} \tag{8.3}$$

In Eq. (8.3), q ranges from 0 to 1 and measures the degree of a point inside a 3D box (Yang et al. 2019). The point closer to the box center has a higher q. q_n and $\bar{q}_n$ respectively represents the q value of the nth point in $\boldsymbol{P}_f$ with $\boldsymbol{B}$ and $\bar{\boldsymbol{B}}$.

Instance Mask Branch. The instance mask prediction branch takes point features $\boldsymbol{F}_p$, the aggregated feature $\boldsymbol{F}_a$ and the predicted bounding box coordinates $\boldsymbol{B}$ as input. Compared with predicting the instance mask solely depending on point features, the supplementary information from the global feature and the instance bounding box offers instance-aware context. The output is instance masks $\boldsymbol{M} \in \mathbb{R}^{N \times \mathbb{I}}$ indicating the probability of points belonging to the detected object, with the last layer being a sigmoid layer. In our experiment, we adopt the Focal Loss to tackle the imbalance between instance points and clutters, which is shown in Eq. (8.4), where α and γ are hyper-parameters (Lin et al. 2017).

$$\mathcal{L}_{mask} = -\frac{1}{N}\sum_{m_i \in M} \alpha\left(1 - m_i\right)^{\gamma} \log\left(m_i\right) \tag{8.4}$$

Semantic Label Branch. Additionally, most point cloud segmentation datasets have point-level semantic annotations, so a straightforward point-level semantic label prediction branch is adopted to exert the backbone to learn useful information from point clouds, and $\mathcal{L}_{sem}$ is a weighted cross-entropy loss as shown in Eq. (8.5). In Eq. (8.5), y_i^n is a binary indicator denoting whether the nth point belongs to class i (totally n_s classes), and correspondingly, p_i^n is the predicted probability. The weight $w_i = \frac{median(r)}{r_i}$ is involved in dealing with the imbalance between classes, where r_i denotes the ratio of points belonging to the ith class and $median(r)$ is the median value of these r_i.

$$\mathcal{L}_{sem} = -\frac{1}{N}\sum_{n=1}^{N}\sum_{i=1}^{n_s} w_i y_i^n \log\left(p_i^n\right) \tag{8.5}$$

A joint multi-task loss function $\mathcal{L}_{joint}$ is used to train this network, as shown in Eq. (8.6).

$$\mathcal{L}_{joint} = \mathcal{L}_{box} + \mathcal{L}_{mask} + \mathcal{L}_{sem} \tag{8.6}$$

8.1.4 Object-centric Instance Segmentation

The output of the pipeline is supposed to be an object-centric road furniture instance segmentation containing semantic and point-level information. Usually, the identical street furniture object may occur in adjacent street-view images with overlapping fields of sight. Therefore, simply accumulating every instance mask $\boldsymbol{M}$ from all images will lead to the accumulation of instance mask prediction errors. In that concern, an effective fusion strategy is used to fuse $\boldsymbol{M}$ from different image frames, which firstly associates frustums containing the same instance. Before we associate the frustums, the predicted bounding boxes $\boldsymbol{B}$ and the point clouds are all transformed into the original projection coordinate system. Given the predicted bounding boxes and the orientation parameters of image frames, we follow the pattern of multi-object tracking (MOT), in which an instance ID is assigned to each frustum. Concretely, we used the MOT method proposed by (Weng et al. 2020), where the 3D bounding box IoU (Intersection over Union) and the semantic class are used as association criteria. Then, the point clouds of each associated frustum group are translated to the object-centric coordinate system (x_o, y_o, z_o), where the coordinate origin is the center of the bounding boxes.

Firstly, a threshold T_{fm} is applied to filter the clutter points, whose $m_f \in \boldsymbol{M}_{frustum}$ is smaller than T_{fm}. Those kept points from different frustums are gathered as $\boldsymbol{Po} \in \mathbb{R}^{N' \times (3+1)}$ with the predicted m_f as the feature channel. Finally, after fusing the instance mask prediction from the associated frustums, a straightforward PointNet++ (Qi et al. 2017) is applied to predict a binary instance segmentation mask. In the instance segmentation network, the overlapping points from different frustums are treated as separated points. And they will be regarded as one instance point if any of them is predicted positive by the network.

8.2 EXPERIMENTS

8.2.1 Dataset Description and Experiment Settings

Street-view images and MLS point clouds from three regions were used in the experiments. For training the neural network, we use the manually-labeled point clouds introduced by (Han, Dong, and Yang 2021), which cover 6.0 km urban roads in Shanghai. Datasets collected from Shanghai and Wuhan are used to validate the proposed method. The Shanghai dataset covers 6.5 km urban roads, 1.3 km of which are manually labeled for quantitative evaluation. The Wuhan dataset covers 3.2 km urban roads, with 1.3 km street furniture annotations.

The instance segmentation of nine classes of street furniture is evaluated, covering typical municipal and transportation assets. Our method outputs the concrete semantic class of traffic signs (127 classes in the training set), and they are categorized into two groups in the point cloud annotations: text signs and warning signs.

After cropping frustums, the point intensity and 3D coordinates are used as the feature channels of P_f. The multi-scale grouping PointNet++ (Qi et al. 2017) is used as the point cloud processing backbone in predicting per-frustum bounding boxes and instance masks and is also used to predict the final object-centric instance masks, with hyper-parameters set according to (Qi et al. 2018). After frustum association, T_{fm} is set to 0.3 to filter the points that are predicted with a very low possibility of being instance points. The mentioned neural networks in the proposed pipeline are trained on an NVIDIA 2080Ti GPU for 50 epochs.

To precisely reflect on the performance of our proposed method, we especially adopt two levels of quantitative evaluation indicators in this study.

Instance-level Metrics. Instance-level recall and precision are listed to show the ratio of correctly inventoried instances at a given point IoU threshold (0.5 in this study). If the point IoU of a predicted and a ground truth (GT) instance is greater than the threshold and the assigned semantic label is correct, it is correctly inventoried (i.e., true positive (TP)). The other predicted instances are regarded as false positives (FP), and the missed GT instances are false negatives (FN). Refer to Eqs. (8.7) and (8.8) for the calculation of instance-level recall and precision.

Point-level Metrics. Our pipeline integrates two point mask prediction neural networks and produces street furniture points, so the point-level indicators (precision, recall, weighted coverage) are adopted. The equations of point recall, precision and IoU are shown in Eqs. (8.7)–(8.9). Point-level TP denotes the correctly predicted instance points of the successfully inventoried instances. FP (False Positive) denotes the wrongly predicted instance points and FN (False Negative) denotes the missed instance points.

$$\text{Recall} = \frac{\text{TP}}{\text{TP} + \text{FN}} \tag{8.7}$$

$$\text{Precision} = \frac{\text{TP}}{\text{TP} + \text{FP}} \tag{8.8}$$

$$\text{IoU} = \frac{\text{TP}}{\text{TP} + \text{FP} + \text{FN}} \tag{8.9}$$

Weighted coverage (wCov) is widely used to evaluate the performance of instance segmentation, which calculates the weighted average instance-wise IoU for ground truth instances $r_i, r_k \in \mathcal{G}$ and their associated predictions $r_j \in \mathcal{P}$, as shown in Eq. (8.10) and (8.11).

$$wCov(\mathcal{G}, \mathcal{P}) = \sum_i w_i \max_j IoU\left(r_i^{\mathcal{G}}, r_j^{\mathcal{P}}\right) \tag{8.10}$$

$$w_i = \frac{\left|r_i^{\mathcal{G}}\right|}{\sum_k \left|r_k^{\mathcal{G}}\right|} \tag{8.11}$$

8.2.2 Qualitative Results

Figure 8.3 illustrates the qualitative experiment results from the global perspective, where the extracted street lamp, traffic light, traffic sign, and trash bin points are overlaid on the raw MLS point clouds, and Figure 8.4 presents close-up results at crossroads. As the experiment results show, the proposed pipeline locates the street furniture and outputs their point clouds in typical urban streets and crossroads, presenting their inventories and geometric features including shapes and sizes for transportation administration. In general, this method processes all types of detected street objects using a universal framework and common parameter settings, and it is valid for all the mentioned classes. This class list is promising to be extended without worrying about excessive labor for designing extra hand-crafted feature descriptors.

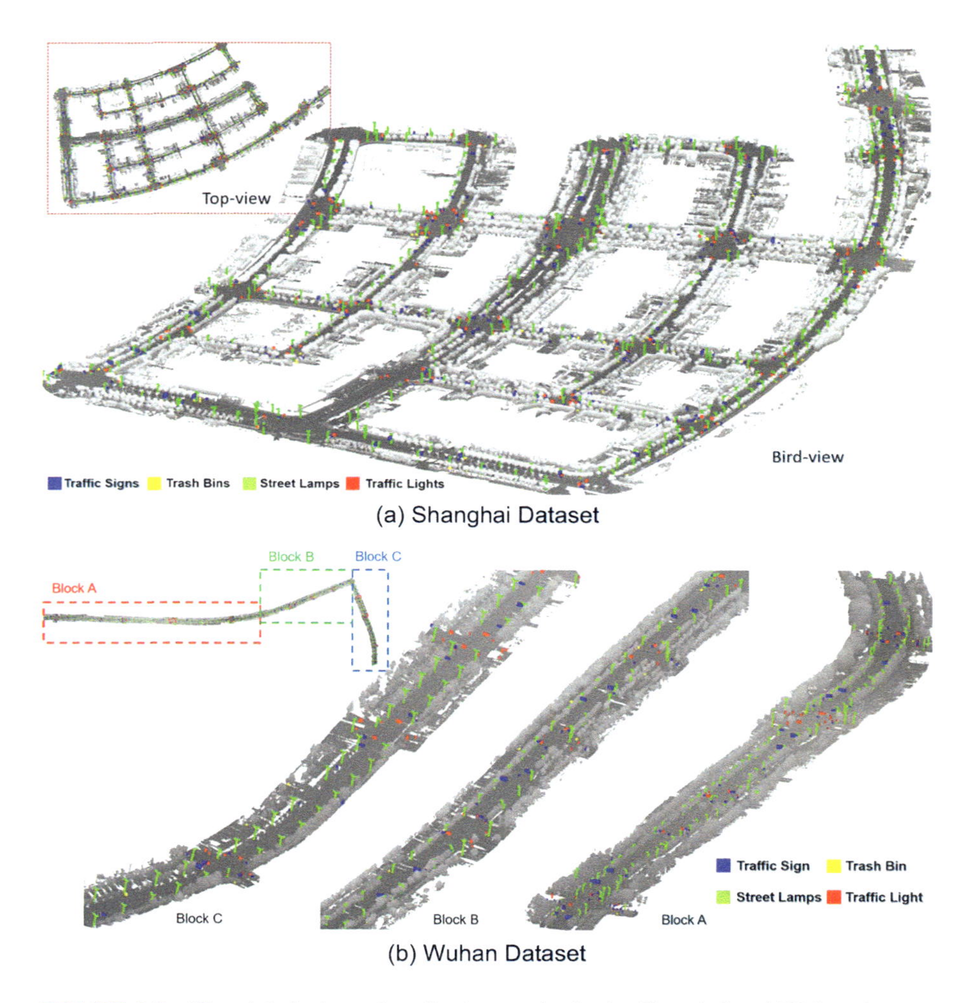

FIGURE 8.3 The global view of qualitative results in the Shanghai and Wuhan datasets (Zhou et al. 2022).

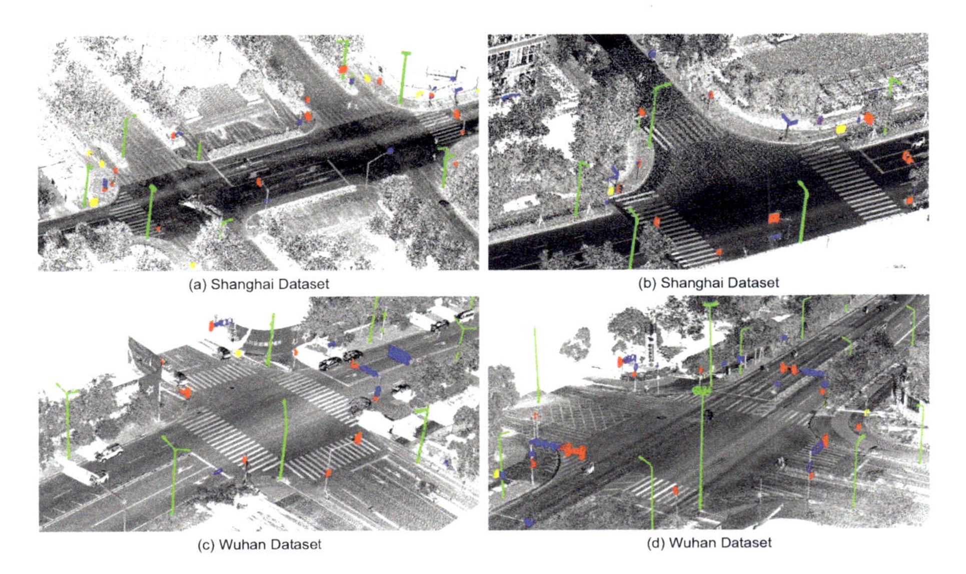

FIGURE 8.4 The close-up view of the qualitative results at crossroads from the Shanghai and Wuhan datasets (Zhou et al. 2022).

The proposed method is capable of segmenting traffic cones, the temporary traffic barriers, supporting the update of HD maps. Besides, it is significant for residents that the fire hydrants are effectively inventoried using the proposed method, considering that they are usually ignored using methods depending solely on point clouds.

In previous studies using street-view imagery or point clouds for street furniture recognition, the dense vegetation in urban streets often caused serious occlusions, especially for lamps. By contrast, since the proposed pipeline merges and integrates the semantic and geometric cues from multiple perspectives by subsequently predicting instance masks in the frustum-centric and the object-centric manners, the occlusions do not cause great interference to the results.

Another characteristic of our method is that it produces component-level results under the semantic guidance from street-view imagery. For example, traffic signs and traffic lights mounted on the same pole are respectively surveyed with detailed traffic sign semantic labels, even the weight or speed limits. This benefits the traffic management authorities by providing detailed distribution and semantic information about road assets for analyzing their rationality in traffic planning and consistency with driving safety.

8.2.3 Quantitative Results

Quantitative evaluation is performed on our annotated street segments, which respectively cover 1.3 km roads in the Shanghai and Wuhan datasets.

Tables 8.1 and 8.2 show the quantitative results in the Shanghai and Wuhan datasets. From the global perspective, the mean instance recall and precision of Wuhan and Shanghai exceed 80.9%, which verifies the effectiveness of our proposed method.

TABLE 8.1
Instance-level and Point-level Quantitative Results in the Shanghai Dataset

Class	Instance-level Statistics (IoU@0.5)					Point-level Statistics (IoU@0.5)		
	Recall/%	Precision/%	TP	FN	FP	Recall/%	Precision/%	wCov/%
Text Signs	78.8	78.8	26	7	7	77.7	81.0	76.4
Traffic Lights	86.8	88.1	59	9	8	86.1	71.5	76.3
Street Lamps	98.0	92.4	97	2	8	95.4	93.6	94.5
Trash Bins	89.7	85.4	35	4	6	89.5	88.3	87.8
Traffic Cones	83.3	75.0	15	3	5	76.0	39.5	69.4
Fire Hydrants	70.0	77.8	7	3	2	53.2	69.2	58.0
Warning Signs	84.4	75.0	27	5	9	62.8	55.6	52.9
Potted Plants	100.0	75.0	3	0	1	91.2	91.1	91.1
Mean	**86.4**	**80.9**				**79.0**	**73.7**	**75.8**

TABLE 8.2
Instance-level and Point-level Quantitative Results in the Wuhan Dataset

Class	Instance-level Statistics (IoU@0.5)					Point-level Statistics (IoU@0.5)		
	Recall/%	Precision/%	TP	FN	FP	Recall/%	Precision/%	wCov/%
Text Signs	85.7	96.0	24	4	1	73.1	84.8	70.4
Traffic Lights	78.9	86.6	71	19	11	86.8	73.6	78.3
Street Lamps	99.0	96.0	96	1	4	90.6	90.4	89.8
Trash Bins	76.2	72.7	16	5	6	63.2	71.6	66.5
Traffic Cones	79.8	89.7	130	33	15	83.4	70.2	76.3
Fire Hydrants	57.1	66.7	4	3	2	52.0	77.3	51.5
Warning Signs	88.7	94.8	55	7	3	89.2	67.7	70.8
Benches	100.0	100.0	4	0	0	100.0	89.1	89.7
Mean	**83.2**	**87.8**				**79.8**	**78.1**	**74.2**

Meanwhile, this universal framework is valid for all the mentioned nine classes. Moreover, the mean metrics of the Shanghai and Wuhan datasets show no evident performance differences. The instance-level recall and precision of street lamps, one of the most basic street furniture, surpass 92.4% in both Shanghai (98.0% and 92.4%) and Wuhan (99.0%, 96.0%), achieving the best results among all categories.

Considering the introduced neural networks in our pipeline are trained with a different MLS dataset in Shanghai. This method can be applied not only in another area in Shanghai, but it is also robust to the scene change in Wuhan with different city landscapes and data collection systems, which guarantees its generalizability.

We notice that for traffic signs, indicators in Wuhan are higher than in Shanghai, and this is because the traffic sign point clouds in Shanghai suffer a more severe loss due to the scanning angle and the reflective properties. On the contrary, the proposed pipeline performs better on trash bins in Shanghai than in Wuhan, since the image object detection misses more trash bin instances. Although the proposed method correctly recalls or locates most traffic cones and fire hydrants, the point-level metrics are evidently lower than other larger traffic assets. The reason is that the point numbers of each instance are limited due to their size, so a few false positive instances may result in a very considerable decrease in the indicators.

8.3 ANALYSIS AND DISCUSSION

8.3.1 Architecture Design Analysis

In this part, we present quantitative results using different pipeline settings to demonstrate the effects of specific modules. The settings are shown in Table 8.3. To facilitate the object-centric evaluation, the frustum association is kept using the distances between centroids of predicted instance points, even if the bounding box branch is removed. The performance comparison is shown in Tables 8.4 and 8.5. Moreover, the effects of the semantic label from images and the point cloud intensity are analyzed.

Multi-view Fusion. In setting A, after frustum association, we eliminate the final object-centric binary mask prediction network. We use the distance between the center of the predicted bounding box B and the center of instance points to measure their consistency. The frustum having the smallest center distance is kept for each object, considered as having the most confident prediction result. Experiment results show an obvious decrease in point-level mRec and mwCov, but the point-level mPrec is not

TABLE 8.3
Settings of the Ablation Study

Setting Group	B. Box Branch	Sem. Labe Branch	Multi-view Fusion
A	x	x	
B			x
C		x	x
D	x		x
Full	x	x	x

TABLE 8.4
The Ablation Study Results in the Shanghai Dataset

Setting Group	mRecall/% (Instance)	mPrecision/% (Instance)	mRecall/% (Point)	mPrecision/% (Point)	mwCov/% (Point)
A	79.2	73.9	62.4	77.7	64.2
B	77.5	71.9	73.7	65.7	72.9
C	78.9	73.8	73.1	67.3	71.4
D	81.8	77.6	77.5	71.8	74.3
Full	86.4	80.9	79.0	73.7	75.8

TABLE 8.5
The Ablation Study Results in the Wuhan Dataset

Setting Group	mRecall/% (Instance)	mPrecision/% (Instance)	mRecall/% (Point)	mPrecision/% (Point)	mwCov/% (Point)
A	78.2	82.6	70.8	73.9	68.2
B	81.6	80.7	78.8	66.8	70.0
C	81.0	86.2	79.4	72.0	71.7
D	81.4	84.0	80.2	74.3	73.2
Full	83.2	87.8	79.8	78.1	74.2

influenced. That is, a large ratio of the instance points predicted by the kept frustum are correct, considering that they produce very confident predictions from the frustum point cloud. However, this leads to the omission of some instance points that are not significant enough in certain frustums, and results in lower point-level coverage and instance-level recall, precision at the IoU threshold. Especially for tough scenes, like lamps partly occluded by trees or fire hydrants partly hidden by vegetation, the fusion-based final mask prediction equips the pipeline with a second chance to compensate for biases in the frustum perspective.

Frustum-centric Neural Network. Different combinations of neural network branches are tested to demonstrate the effects of the bounding box branch and the semantic label branch. Generally, compared with solely predicting instance masks, both these two branches show their effects in promoting the inventory performance. They are mutually reinforcing in this multi-task setting. The bounding box branch, especially, provides an evident boost. This validates our intention to add instance-aware context to strengthen the instance mask prediction.

Image Class Labels and Point Intensity. The effects of image class labels and the point intensity are also investigated. The experiments show that image semantic labels are of great importance, especially when the frustum contains an instance of another class, which often occurs at a crossroads. The effect of point intensity is not very prominent, which is good for portability using other MLS systems. But in Shanghai Dataset, the recall and precision of text signs drops from 78.8%, 78.8% to 69.7%, 57.6% without the point intensity. This is because text sign frustums sometimes contain the building façade, showing similar planar geometry features, but their intensity features are usually very different.

8.3.2 Comparative Studies

In this part, Frustum-PointNet (Qi et al. 2018) is discussed as a baseline framework to further illustrate our contributions. Frustum-PointNet is implemented and trained on our experiment dataset, and the quantitative comparison is shown in Table 8.6. To facilitate the instance-level comparison, the same frustum association strategy is used. Frustum-PointNet is an inspiring method integrating images and point clouds for 3D object detection, producing instance masks and amodal bounding boxes. It firstly predicts the point-level instance mask and then uses it for bounding box regression. Therefore, the bounding box regression is heavily dependent on the results of instance segmentation, and the instance segmentation lacks instance-aware information. Comparatively, the proposed method enhances the instance context information in per-frustum instance mask prediction by involving the global feature and the bounding box supervision as assistance. The instance-level mean recall and precision increased from 70.8%, 61.7% to 86.4%, 80.9% in Shanghai, from 76.4%, 77.3% to 83.2%, 87.8% in Wuhan. The increase in the instance segmentation performance verifies the effectiveness of our design.

8.3.3 Error Analysis

Although the introduced method achieved street object instance segmentation in the experiments, there are some aspects that need attention in future applications and improvements. We have analyzed three typical types of errors in our experiments.

Object Detection Error. The proposed method is based on the guidance of image object detection, and hence is influenced by the quality of 2D proposals. Typical object detection errors mainly constitute semantic label errors and object omissions, and they account for most errors for street lamps and trash bins. *Instance Mask Error*. Based on the 2D proposals, our method achieved street furniture point cloud instance segmentation. Instance mask errors refer to those caused by incorrect mask prediction from the neural networks. For traffic signs and traffic lights, the instance mask prediction errors are comparatively more considerable. Also, for fire hydrants and traffic cones, instance mask prediction is more challenging because of the great point number imbalance between instance points and background. *Object Missing in Point Clouds*. Although some objects are detected in the images, the corresponding point clouds are not successfully collected, which takes up a minor part of the errors. For example, due to the scanning range and the reflective property of the traffic sign, the traffic sign shows very few point numbers, which is typical in some MLS datasets.

TABLE 8.6
The Quantitative Comparison with Frustum-PointNet

		Instance-level Statistics (IoU@0.5)		Point-level Statistics (IoU@0.5)		
Dataset	**Method**	**mRecall/%**	**mPrecision/%**	**mRecall/%**	**mPrecision/%**	**mwCov/%**
Shanghai	F-PointNet	70.8	61.7	59.1	69.3	60.5
	Ours	86.4	80.9	79.0	73.7	75.8
Wuhan	F-PointNet	76.4	77.3	66.3	71.2	62.6
	Ours	83.2	87.8	79.8	78.1	74.2

8.4 CONCLUSION

This chapter presents a method for street scene point cloud instance segmentation incorporating street-view imagery. The proposed method is thoroughly introduced, which contains three major steps: (1) frustum cropping; (2) per-frustum instance mask and 3D bounding box prediction; (3) object-centric instance segmentation. This pipeline is validated by experiments in Shanghai and Wuhan datasets, producing component-level street furniture instance segmentation with concrete semantic labels and instance point clouds. Experiments report the mean instance recall and precision respectively reach 86.4%, 80.9% and 83.2%, 87.8% in Shanghai and Wuhan, and the point-level mean recall, precision, wCov all exceed 73.7%, meeting the demand of urban administration and outperforming previous studies.

REFERENCES

Han, Xu, Zhen Dong and Bisheng Yang. 2021. "A point-based deep learning network for semantic segmentation of MLS point clouds." *ISPRS Journal of Photogrammetry and Remote Sensing* 175:199–214. doi:10.1016/j.isprsjprs.2021.03.001

Lin, Tsung-Yi, Priya Goyal, Ross Girshick, Kaiming He, and Piotr Dollár. 2017. "Focal loss for dense object detection." In *Proceedings of the IEEE international conference on computer vision*:2980–2988. doi:10.1109/TPAMI.2018.2858826

Qi, Charles Ruizhongtai, Li Yi, Hao Su, and Leonidas J. Guibas. 2017. "Pointnet++: Deep hierarchical feature learning on point sets in a metric space." *Advances in Neural Information Processing Systems* 30. doi:10.48550/arXiv.1706.02413

Qi, Charles Ruizhongtai, Wei Liu, Chenxia Wu, Hao Su, and Leonidas J. Guibas. 2018. "Frustum pointnets for 3D object detection from rgb-d data." In *Proceedings of the IEEE conference on computer vision and pattern recognition*: 918–927. doi:10.1109/CVPR.2018.00102

Shao, Shuai, Zeming Li, Tianyuan Zhang, Chao Peng, Gang Yu, Xiangyu Zhang, Jing Li, and Jian Sun. 2019. "Objects 365: A large-scale, high-quality dataset for object detection." In *Proceedings of the IEEE/CVF international conference on computer vision*: 8430–8439. doi:10.1109/ICCV.2019.00852

Weng, Xinshuo, Jianren Wang, David Held, and Kris Kitani. 2020. "3D multi-object tracking: A baseline and new evaluation metrics." In *2020 IEEE/RSJ International Conference on Intelligent Robots and Systems (IROS)*:10359–10366. doi:10.1109/IROS45743.2020.9341164

Yang, Bo, Jianan Wang, Ronald Clark, Qingyong Hu, Sen Wang, Andrew Markham, and Niki Trigoni. 2019. "Learning object bounding boxes for 3D instance segmentation on point clouds." *Advances in Neural Information Processing Systems* 32. doi:10.48550/arXiv.1906.01140

Zhou, Yuzhou, Xu Han, Mingjun Peng, Haiting Li, Bo Yang, Zhen Dong, and Bisheng Yang. 2022. "Street-view imagery guided street furniture inventory from mobile laser scanning point clouds." *ISPRS Journal of Photogrammetry and Remote Sensing* 189:63–77. doi:10.1016/j.isprsjprs.2022.04.023

Zhu, Zhe, Dun Liang, Songhai Zhang, Xiaolei Huang, Baoli Li, and Shimin Hu. 2016. "Traffic-sign detection and classification in the wild." In *Proceedings of the IEEE conference on computer vision and pattern recognition*: 2110–2118. doi:10.1109/CVPR.2016.232

Part IV

Modeling and Analysis of Point Cloud

The generation of high-quality 3D digital elevation models (DEMs), building models, and road models are vital applications of ubiquitous point clouds. Although LiDAR offers rapid data collection, high accuracy, and the ability to penetrate vegetation, it often produces disorganized, redundant data that lacks crucial semantic, structural, and spatial information. This complicates its application across diverse domains and affects the quality of feature extraction and model reconstruction. This part highlights the significant contributions and insights from three interconnected aspects of point cloud modeling and analysis: 3D terrain modeling, building structural modeling, and road modeling, emphasizing their collective impact on advancing geospatial technologies.

This part first introduces a two-step adaptive extraction method for ground points and breaklines from point clouds, an essential fundamental step in generating DEMs. Traditional filtering techniques, categorized into methods based on point entities and those based on segment entities, often struggle to effectively process airborne LiDAR point clouds across diverse and complex terrains. Moreover, combining different point entity-based filtering methods may not effectively differentiate between features such as cliffs and buildings within a local area, and existing methods largely overlook the potential benefits of integrating different entity-based methods. To address these limitations, we introduce a robust extraction method that utilizes a fusion of two distinct filtering techniques tailored to different entities, enabling the detection of multiple types of breaklines from the extracted ground points. The proposed method begins by organizing the points into segments and individual units, which are then subjected to segment-based and multi-scale morphological filtering. Multi-scale morphological filtering specifically targets the removal of amorphous objects from individual points to mitigate the impact of the largest scale on the filtering results. Following this, the method extracts breaklines from the ground points, establishing a solid foundation for creating high-quality DEMs. Experimental results confirm that this method not only robustly extracts ground points but also effectively preserves the integrity of the breaklines.

Additionally, reconstructing 3D building models from point clouds is crucial for various smart city applications like spatial queries, analysis, and visualization. We present a building reconstruction approach from point clouds. The process of automatically reconstructing 3D buildings from point clouds is notably challenging due to the intricate nature of building structures and the constraints of current reconstruction algorithms. Existing methods fall into three categories: data-driven, model-driven, and hybrid-driven, each grappling with issues such as sensitivity to incomplete data caused by occlusions, shadows, and missing information, limitations imposed by pre-defined primitives for roof structures, and challenges in accurately extracting roof planes. To overcome these limitations, we introduce a novel approach that utilizes a hierarchical topology tree to enhance the recognition and interpretation of 3D building models from raw point clouds. Specifically, a Roof Attribute Graph (RAG) method is proposed to articulate the decomposition and topological relationships within complex roof structures. Furthermore, top-down decomposition and bottom-up refinement processes are proposed to reconstruct roof parts according to the Gestalt laws, generating a complete structural model with a hierarchical topological tree. Experimental results, show that the proposed method can be used to model 3D building roofs with high quality results as demonstrated by the completeness and correctness metrics presented in this part.

Another pivotal application of point clouds is the HD map generation and road construction, leverage mobile laser scanning (MLS) for detecting and classifying road boundaries and markings. Despite progress, most studies still primarily focus on classifying road points, with significant gaps in 3D road boundary vectorization and handling worn or incomplete road markings. To address these challenges, we introduce two innovative approaches for road feature extraction and vectorization from MLS data. Firstly, we introduce a supervoxel generation method for boundary extraction, which not only preserves fine borders but also enhances computational efficiency. Then a lightweight but precise 3D vectorization technique using adaptive cubic Bezier curves and a standard Kalman filter are adopted. These processes result in continuous road boundary segments that are further refined to provide accurate road geometric parameters. Secondly, for road marking extraction and vectorization, we employ a two-stage coarse-to-fine object detection and localization strategy. This method starts with a general object detection network to identify road marking bounding boxes, followed by a shape matching operator that refines positions, orientations, and scales based on standard geometric and radiometric properties. Experimental results demonstrate that our proposed methods achieve robust performance across various road shapes and conditions, and effectively handle worn and incomplete markings.

In summary, this part emphasizes innovative approaches in point cloud modeling, specifically addressing key challenges in 3D terrain modeling, building structural modeling, and road modeling. These methodologies enhance the extraction and vectorization of features from complex environments, providing improved accuracy and efficiency in the creation of digital elevation models, 3D building structures, and detailed road maps. This collective advancement in geospatial technologies underscores a significant step forward in managing and interpreting the vast data provided by LiDAR and other scanning technologies.

9 3D Terrain Modeling

9.1 METHODOLOGY

In this section, an adaptive point cloud filtering method is proposed, which integrates the point cloud segmentation filtering and the multi-scale morphological filtering benefiting processing breaklines. Figure 9.1 displays the workflow of the adaptive point cloud filtering method, which includes three main steps. First, the pseudo-grids with a certain size are generated from airborne LiDAR point clouds and the point clouds are divided into gridded points and other point clouds. Secondly, the gridded points are adaptively partitioned into two parts: the segments and the discrete independent point cloud set. The ground grid point cloud is extracted using the point cloud partitioning filtering method and the multi-scale morphological filtering method. Thirdly, the provisional DEM is generated based on the ground grid point cloud. This provisional DEM serves as a foundation to extract the ground point cloud from other point clouds that were separated in the first step. This process allows for the identification of all ground point clouds that meet the conditions in the study area. Finally, in order to generate high-quality DEM, the breaklines are extracted from the final ground points.

9.1.1 Generation of Pseudo-grids

In point cloud filtering, the data quality of the point cloud has a great impact on the results and computational efficiency. If the density of the point cloud is too high, it requires a lot of computational capacity to compute the local features of the point cloud and perform the filter process; if the density of the point cloud is too low or unevenly distributed, it will seriously affect the accuracy of the local features of the point cloud, thus affecting the results of the point cloud filtering; the missing data area may cause a discontinuity in the distribution of the ground point cloud, leading to the weakening of the neighborhood spatial relationship between the point clouds, thus increasing the probability of point cloud filtering errors. For this reason, according to Cho et al. (2004), a pseudo-grids generation technique is used to process the original point cloud to make the point cloud distribution as uniform as possible and to decrease the data blanks, as shown in Figure 9.2.

9.1.2 Ground Points Extraction from Grid Points

Ground points extraction mainly involves two steps. First, the point cloud is adaptively partitioned by the point cloud segmentation method to divide the grid point cloud into two types: segments and individual point cloud sets; then, the two types of point cloud are processed by the point cloud segmentation filtering

DOI: 10.1201/9781003486060-13

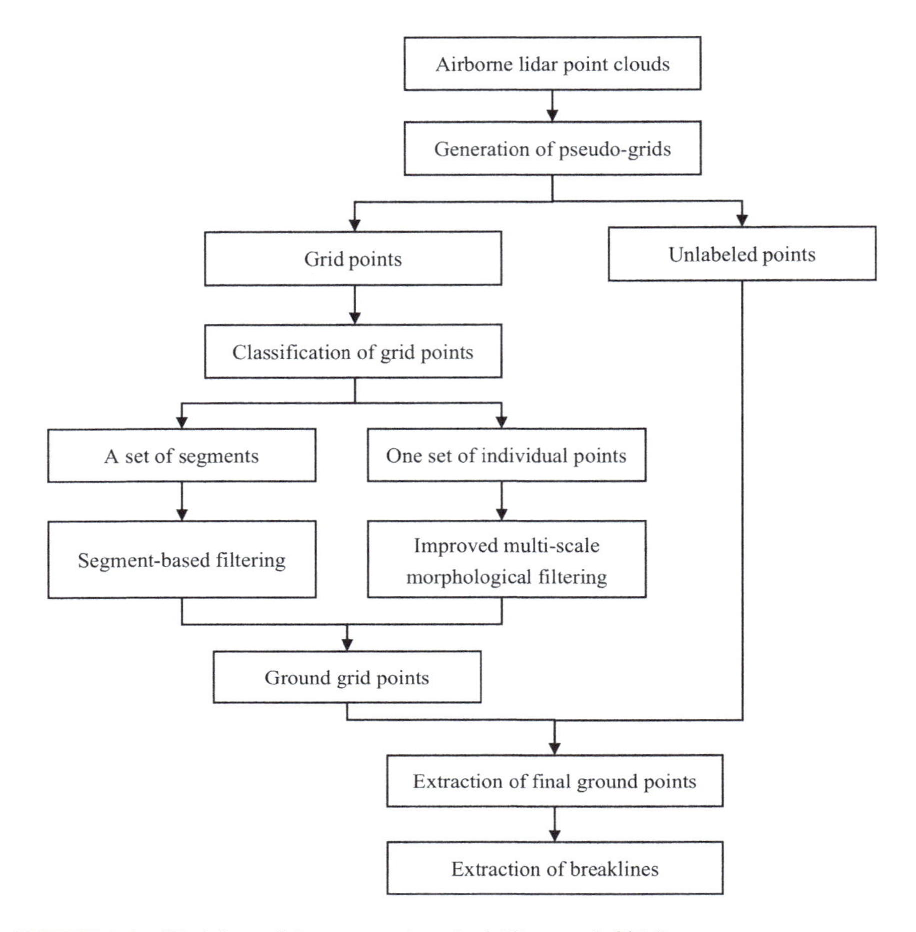

FIGURE 9.1 Workflow of the proposed method (Yang et al. 2016).

and multi-scale morphological filtering methods, respectively, to divide the grid points into ground points and non-ground points.

9.1.2.1 Classification of Grid Points

Different types of regions should be expressed using different primitives; smooth regions are suitable to be represented as segment primitives with regionality, and point primitives should be used for regions with drastic changes, so that the differences between ground and non-ground point clouds can be described more robustly. In order to realize different primitives for different regions, point cloud segmentation based on smoothing constraints is utilized to divide the point cloud into smooth and gentle regions and regions with drastic elevation changes. Based on the method of Tóvári and Pfeifer (2005), smoothing constraint-based point cloud segmentation uses a region growing method to group neighboring point clouds with similar normal vectors and plane fitting residuals into non-overlapping segments $S = \{S_1, S_2, \ldots, S_{n-1}, S_n\}$, where S_i represents a certain segment and n is the total number of segments. Figure 9.2e shows the segmentation result corresponding to the grid point cloud in

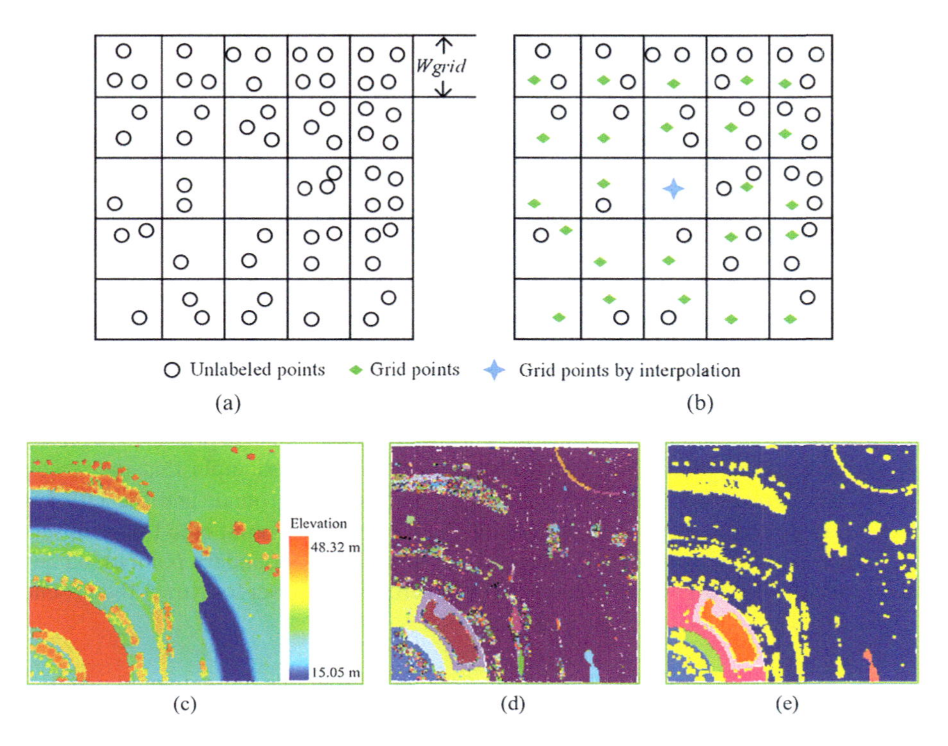

FIGURE 9.2 Generation of pseudo-grids and classification of grid points. (a) Unlabeled points in pseudo-grids. (b) The determination of grid points. (d) Grid points. (e) Segmentation of grid points, each segment is dotted in one color. (f) The result of classification for the grid points, where the segments are dotted in different colors, and the individual points are dotted in yellow (Yang et al. 2016).

Figure 9.2d. The segment size indicates the elevation change of the point cloud in the region. Larger segment slices are typically found in areas with slight changes; smaller segments are found in areas with drastic changes, and the point clouds within the segments may come from different target surfaces, i.e., the sub-segmentation problem. Figure 9.2e shows that smooth regions, such as the ground and buildings, have relatively large segments, while regions with vegetation have broken segmentation effects and relatively small segments. Therefore, adaptive partitioning is performed based on the number of segmented point clouds. First, all segments are traversed, and the number of point clouds in one segment is compared to the set threshold N_t. When the number of point clouds exceeds the threshold, they are represented as segment primitives. If not, they are removed from the set of segments and the point clouds within them are represented as point primitives. This results in the point cloud being divided into a collection of segments and a collection of individual point clouds, as illustrated in Figure 9.2f. After classification, two classes are generated: a set of segments consisting of $S_{\text{homo}} = \{S_1, \ldots, S_i, \ldots, S_k\}$, where S_i denotes a segment in the set, and k is the total number of remaining segments; and one set of individual points $P_{\text{heter}} = \{p_1, \ldots, p_i, \ldots, p_m\}$, where p_i represents a point in the set, and m is the number of the discrete independent point cloud in the set.

9.1.2.2 Identification of Ground Points

After adaptive partitioning, point cloud segmentation filtering and multi-scale morphological filtering are performed together for integrated processing to improve the algorithm's adaptability in complex scenes and diverse terrain regions. In particular, the point cloud segmentation filtering method classifies the segments S_{homo} into ground segments or non-ground segments, while the multi-scale morphological filtering method classifies the points in P_{heter} are labeled as ground or non-ground.

9.1.2.2.1 Segment-based Filtering

Segment-based point cloud filtering for segmentation is mainly based on the existence of a certain elevation difference between the boundary point cloud of non-ground and the ground point cloud to remove the non-ground segments. Therefore, firstly, the initial ground point cloud is extracted by the principle that the lowest point in a certain neighborhood is the seed point of the ground, and the DEM is generated as a temporary ground model; then, the boundary point cloud of each segment is extracted and the elevation difference between the boundary point cloud and the provisional DEM is calculated; finally, a certain rule is set to extract the ground segments. The detailed process is as follows:

Step 1 Provisional DEM Generation. Ground seed points are extracted using the moving window method. A window with a length and width of W_{seed} is moved along the x and y directions with a step of half of W_{seed}, respectively. The lowest point in the moving window is obtained as the ground seed point. Theoretically, W_{seed} should be larger than the object with the maximum size. However, accurately determining the largest feature can be challenging, leading to a small number of erroneous ground seed points. To eliminate these errors, it is necessary to extract the slopes or other features between the ground seed points. The surface containing ground seed points is labeled as a ground seed surface because the point cloud within the surface is homogeneous. Subsequently, the extracted ground seed points are utilized to create a provisional DEM with a resolution of W_{grid} using the Inverse Distance Weighting interpolation (IDW) method, which is denoted as Z_{rtm}.

Step 2 Segment boundary point cloud extraction and elevation difference calculation. All unclassified segments were traversed and the boundary point cloud of each segment was extracted using the α-shape method Edelsbrunner and Mücke (1994). The boundary point cloud of all segments, denoted as ND = $\{\text{ND}_1, \ldots, \text{ND}_i, \ldots., \text{ND}_k\}$, where k denotes the number of the boundary, $\text{ND}_i = \{p_1, \ldots, p_i, \ldots, p_{ik}\}$ denotes the boundary point cloud for a particular segment, pi is a boundary point in ND_i, ik is the number of ND_i boundary point clouds. Then, the elevation difference between each boundary point cloud in provisional DEM is calculated. the set of elevation differences corresponding to ND is denoted as ED = $\{\text{ED}_1,$

…, ED_i, …., ED_k}, and the set of elevation differences corresponding to ND_i is denoted as $ED_1 = \{e_1, \ldots, e_i, \ldots, e_{ik}\}$.

Step 3 Ground Segment Extraction. Considering the observation that most of the boundary points of the non-ground segments have significant elevation differences from the provisional DEM (Chen, 2009), Equation (11.1) is applied to classify the unclassified portions into ground and non-ground.

$$\begin{cases} \mathrm{GS} = \{S_i \in S_{\mathrm{homo}} \mid \max\left(\mathrm{ED}_i\right)\left\langle h_1 \right|\mid \mathrm{Num}(\mathrm{ED}_i < h_2) > 50\% * \mathrm{Num}\left(\mathrm{ND}_i\right) \\ \mathrm{OS} = \left\{S_i \in S_{\mathrm{homo}} \ \& \ S_i \notin \mathrm{GS}\right\} \end{cases} \tag{9.1}$$

Here, GS denotes the set of ground segments, OS represents the set of non-ground segments, h_1 and h_2 are elevation difference thresholds, max(ED_i) is a function that calculates the maximum value within ED_i Num($ED_i < h_2$) counts the number of point clouds in ED_i that are greater than h_2, and Num (ND_i) denotes the quantity of boundary point clouds in ND_i. Empirically, h_1 is set between 0.5 to 0.8 meters, and h_2 is set between 0.2 to 0.4 meters. The threshold h_2 primarily addresses instances where portions of a ground segment are attached to the ground, which is less than h_1.

9.1.2.2.2 Improved Multi-scale Morphological Filtering

For sets of individual points, a multi-scale morphological filtering method is employed to remove non-ground points. After adaptive partitioning, although large man-made structures may not be present, dense vegetation may remain, necessitating a relatively large maximum scale for the multi-scale morphological filter and thus leading to substantial cut-off issues. Additionally, determining the extent of individual dense vegetation poses a challenge for setting the maximum scale for multi-scale morphological filtering. The approach is as follows:

Step 1 Initially setting a smaller scale W_{scale}; then using morphological opening combined with point cloud segmentation filtering to eliminate objects larger than W_{scale}.

Step 2 Applying morphological opening across the area covered by the set of individual points with a window of size W_{scale} to smooth uneven target surfaces within the area. Typically, W_{scale} is set between 5 m and 8 m. For instance, this results in a noticeably smoother surface of trees, as demonstrated in the comparison of vegetation cross-sections before and after morphological opening.

Step 3 Segmenting the smoothed set of individual points and classifying the results to remove non-ground sets.

Step 4 Following the removal of large, unpenetrated vegetation, the elevations of unclassified points within the area are restored to their elevations before the opening operation. The multi-scale morphological filtering method Pingel et al. (2013) is then applied to extract ground grid points, using the provisional DEM for elevation correction.

9.1.3 Final Ground Points Extraction

After the generation of pseudo-grids processing of the point cloud, the point cloud is divided into gridded points P_{G} and other point clouds P_{NG}. The point cloud segmentation filtering and improved multi-scale morphological filtering are used to divide the gridded point cloud into ground and non-ground points, while the other point clouds P_{NG} remain unclassified. Therefore, it is necessary to divide the other point clouds P_{NG} into ground point clouds and non-ground point clouds, so as to extract all the ground point clouds that meet the requirements in the study area. First, the provisional DEM is updated and re-interpolated a DEM with a spatial resolution of W_{grid} using the extracted ground grid points. Then, according to the results of Chen et al. (2007) and Pingel et al. (2013) processing method, the distance from the unclassified point cloud to the temporary DEM and the corresponding terrain slope are considered, and the length of the projection of the distance from the unclassified point cloud to the temporary DEM in the direction of the normal vector h is calculated (as shown in Eq. (9.2)). h (as in Eq. (9.3)) and divide the unclassified point cloud into ground and non-ground point clouds according to Eq. (9.3).

$$h = hv * \cos(s) \tag{9.2}$$

where h_v is the distance from the unclassified point cloud to the temporary DEM along the direction of the plumb line; s is the terrain corresponding to the unclassified point cloud slope, which can be calculated from the temporary DEM.

$$P_{\mathrm{Label}} = \begin{cases} 2, & \text{if } h < h_3 \\ 0, & \text{else} \end{cases} \tag{9.3}$$

where P_{Lable} is the classification result, in which 2 and 0 indicate the terrain and non-terrain respectively. h_3 is the distance threshold, which is related to the point coordinate's accuracy and density, and is usually set between 0.2 m and 0.4 m.

9.1.4 Breaklines Extraction

The extracted ground points are firstly interpolated as a range image to extract the approximate breaklines using a LoG operator (Brügelmann 2000). Since the representation of the range image in the breaklines region has some mistakes, there is a positional bias between the real breaklines and the approximate breaklines. Therefore, the approximate breaklines extraction is buffered by a certain distance as a candidate region for the breaklines, where the buffer distance is usually set to 2~5 m. Next, the point cloud of the candidate region for the breaklines is split into a set of overlapping blocks. Wherein, the size of the block is W_{len}, which can be set to be about 10 m. The block size can be set as about 10 m. In addition, in order to facilitate the connection of fracture lines between neighboring blocks, there needs to be a 30%~50% overlapping between neighboring blocks. The result of the blocks is shown in Figure 9.3. According to the existence of elevation, slope and other sudden changes in the breaklines area, the breaklines are extracted by using the point cloud segmentation in each breaklines block.

Step 1, the point cloud of the candidate region is segmented by the region growing method, as shown in the sub-figure ① in Figure 9.3f.
Step 2, the boundaries of the point cloud within a block and the boundaries of each segment are extracted using the α-shape algorithm, as shown in sub-figures ② and ③ of Figure 9.3f, respectively.
Step 3, the breaklines are extracted using the topological relationship analysis among the boundaries. Due to the abrupt terrain change in the region of the fracture line, the intra-block point cloud is divided into multiple segments, and the fracture line should be located on the boundary of the segments. After analyzing the boundary of the each segment the whole block in a topological relationship, the overlapping boundary points between the two are removed. The remaining boundaries are the breaklines within the block, as shown in the subplot ④ in Figure 9.3f.

After step (3), the breaklines within each point cloud block are extracted, as shown in Figure 9.3g. However, the same breaklines may be distributed in different blocks in the candidate region due to the chunking of the point cloud, which needs to be connected. Therefore, the point cloud of the candidate region is divided into blocks. The exact breaklines may be distributed in different blocks, which need to be connected. In the overlapping region between neighboring blocks, the sub-breaklines from the same breaklines in the neighboring blocks should have some overlap, so that the breaklines in the neighboring blocks can be connected and merged to produce complete breaklines, as shown in Figure 9.3h.

A spline curve is used to fit the breaklines to obtain smoother breaklines as shown in Figure 9.3i.

9.2 EXPERIMENTS

To validate the effectiveness and reliability of the algorithm, test datasets provided by the International Society for Photogrammetry and Remote Sensing (ISPRS) are chosen along with another large-scale scene dataset. Additionally, Ground Truth is provided for each dataset for quantitative evaluation and analysis of the algorithm.

The ISPRS dataset comprises 15 sample datasets, including nine urban and six rural areas. The adaptive point cloud filtering method is applied to the 15 datasets using the same parameter values, as shown in Table 9.1. The filtering results are shown in Figure 9.4. It can be observed that the result shows good performance in preserving terrain features.

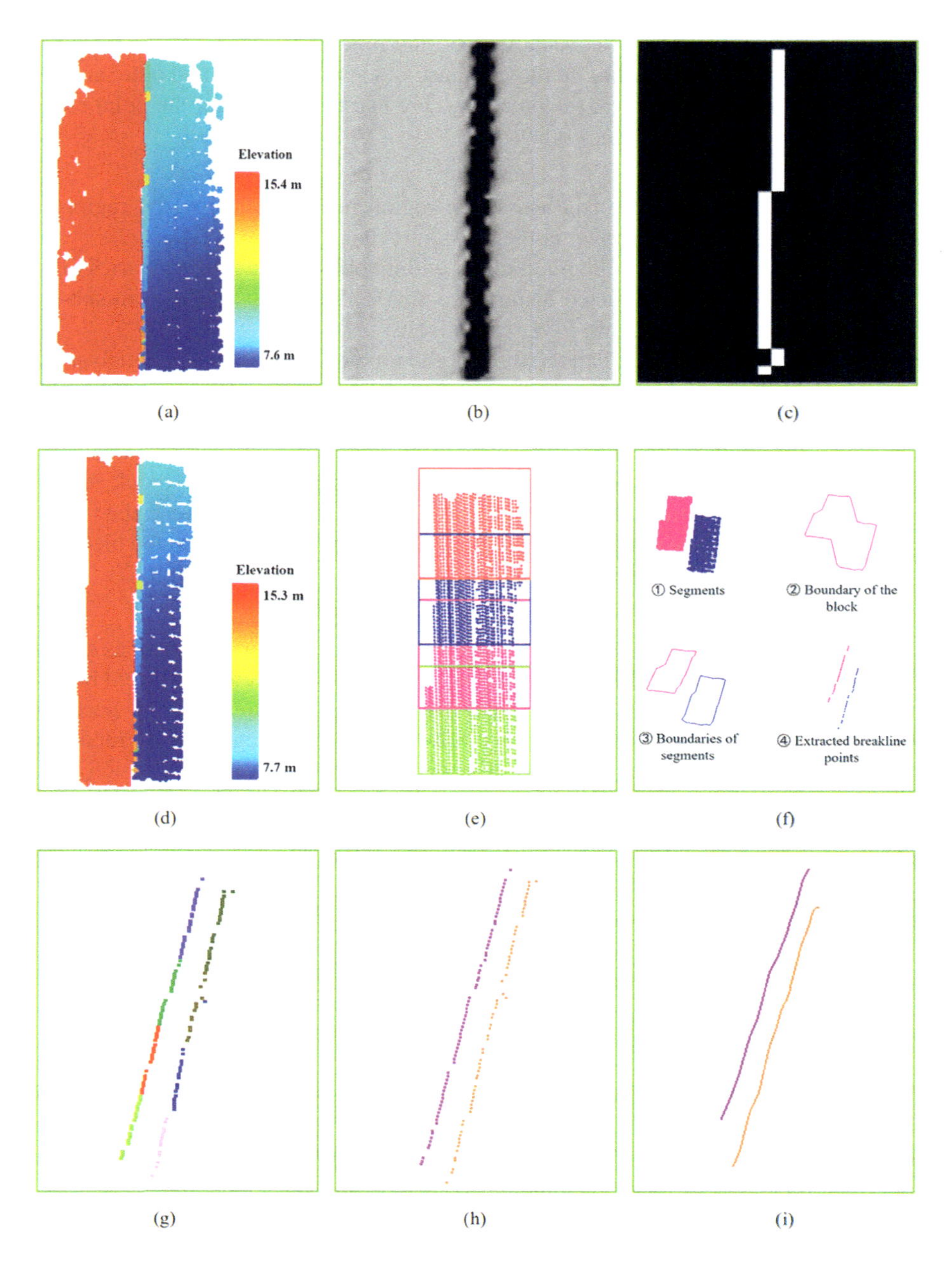

FIGURE 9.3 An example of extracting breaklines from ground points. (a) The extracted ground points. (b) The generated range image. (c) The approximate 2D breakline. (d) The candidate points around the approximate breakline. (e) Generation of the overlapping blocks. (f) The process of the breaklines extraction from one block. (g) The extracted breaklines of all blocks. (h) The linked breaklines across blocks. (i) The fitted breaklines. (Yang et al. 2016).

TABLE 9.1
Parameters Settings for the 15 Sample Datasets

Parameters	W_{grid} (m)	N_t (pts)	W_{seed} (m)	h_1 (m)	h_2	W_{scale} (m)	h_3 (m)
Values	1.0	10	25	0.5	0.3	6.0	0.3

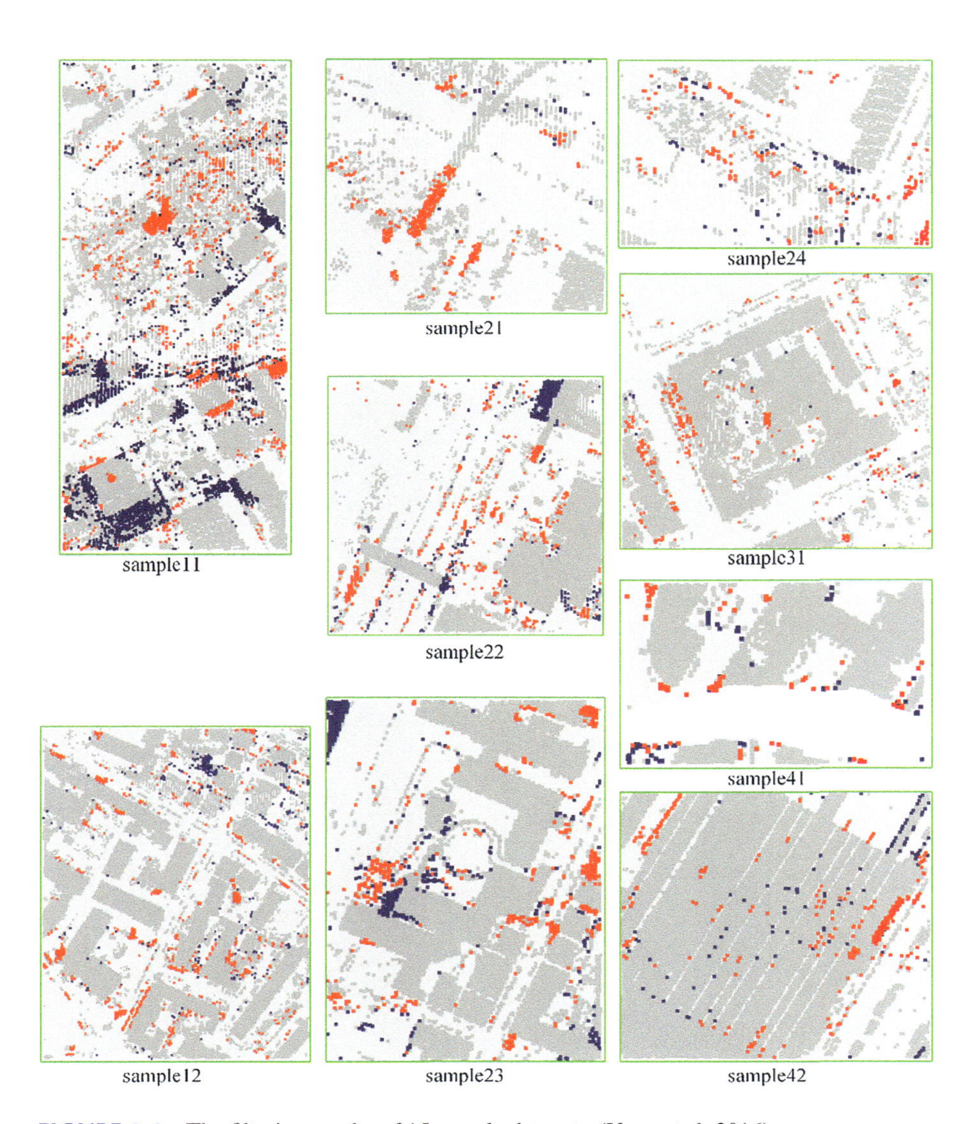

FIGURE 9.4 The filtering results of 15 sample datasets. (Yang et al. 2016).

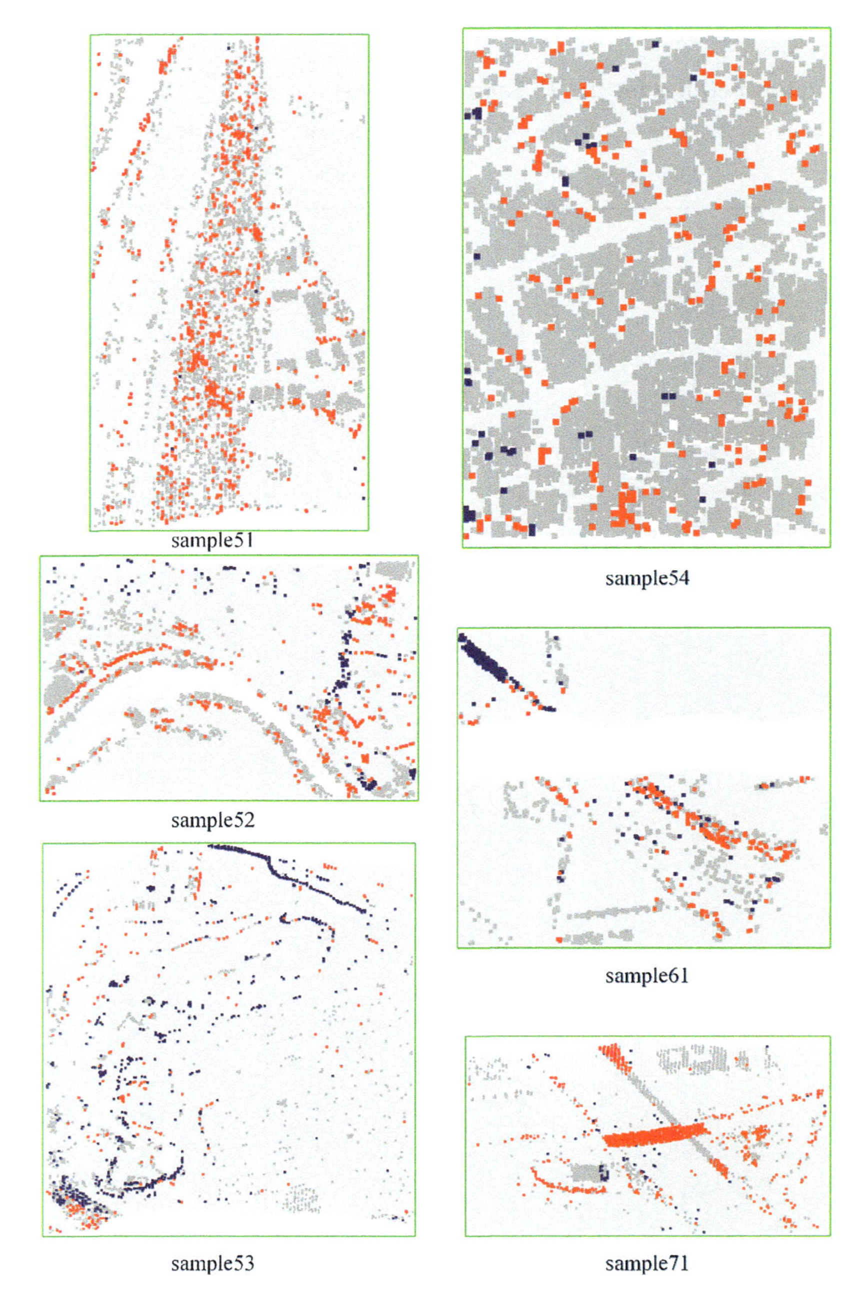

FIGURE 9.4 (Continued)

9.3 ANALYSIS AND DISCUSSION

9.3.1 Performance Evaluation and Comparison

To quantitatively evaluate the algorithm's performance, the type I error (T.I), the type II error (T.II), the total error (T.E), and the kappa coefficient (k) (Sithole and Vosselman (2004)) are calculated for each dataset. The evaluation results of the 15 datasets are then statistically analyzed, as shown in Table 9.2. From Table 9.2, the average total error and average kappa coefficient of the 15 datasets are 3.51% and 87.40%, respectively, with median total error and kappa coefficient values of 3.11% and 91%, respectively. This demonstrates the algorithm's overall good performance, which is able to minimize Type I error while controlling Type II error as much as possible. Additionally, Sample 61 has the lowest total error, while Sample 42 has the highest kappa coefficient. Conversely, Sample 11 has the highest total error, and Sample 53 has the lowest kappa coefficient. The main reasons for the discrepancies in results for Samples 11 and 53 are complex terrain changes and insufficient non-ground point clouds, respectively.

Furthermore, in terms of Type I error, 80% of the datasets are below 3.00%, indicating the algorithm's robustness in various conditions. For example, Sample 53, characterized by numerous terrain abrupt changes, still has a small Type I error of only 2.5%, and most of the omitted ground point clouds are distributed in the cliffs of the quarry, as shown in Figure 9.4b, which does not have much effect on the quality of the DEM generation. However, in terms of Type II error, the algorithm's performance is less satisfactory, with 60% of the datasets exceeding 5.00%. This is mainly due to the small proportion of non-ground points in many datasets, leading to few commission points causing a large type II error. Additionally, errors in point cloud undersegmentation and extraction of low vegetation contribute to the increase in Type II error.

However, compared to Type I error points, post-editing of Type II error points is relatively easier. This is because Type II error point clouds often have significant visible elevation differences with the surrounding ground points, such as the misclassified building point cloud in Sample 11 (Figure 9.4a) and the misclassified bridge point cloud in Sample 71 (Figure 9.4b). Furthermore, although there is a large discrepancy in point spacing between the nine urban area datasets and the six rural area ones, the average total errors are relatively small, with an average total error of 3.76% for urban areas and 3.13% for rural areas. Additionally, a performance comparison of this algorithm with three other algorithms proposed by Mongus et al. (2014), Pingel et al. (2013), Hu et al. (2014)) is conducted using the same parameter values on 15 datasets. The comparison results of total errors and kappa coefficients are presented in Table 9.2, indicating that among the 15 datasets, this algorithm performs best in both total errors and kappa coefficients for eight datasets. These datasets include Sample 12, Sample 23, Sample 24, Sample 31, Sample 41, Sample 52, Sample 53, and Sample 61. These eight datasets present various filtering challenges such as diverse scenes, varied terrains, data voids, density variations, and low densities. In the remaining seven datasets, poor algorithm performance is attributed to reasons such as overly coarse temporary DEMs, which fail to capture frequent terrain changes in Sample 11 and Sample 22, resulting in missed ground point extraction; in Sample 21, Sample 22, and Sample 71, segmentation issues lead to misclassification of some surface-attached objects as ground point clouds; and the presence of low-height vegetation and sparse point cloud density also affects algorithm performance to some extent, as observed in Sample 51 and Sample 54.

TABLE 9.2
Performance Evaluation and Comparison

	T.I (%)	T.II (%)	T.E (%)				k (%)			
Sample	Proposed method		Mongus et al. (2014)	Pingel et al. (2013)	Hu et al. (2014)	Proposed method	Mongus et al. (2014)	Pingel et al. (2013)	Hu et al. (2014)	Proposed method
11	11.69	8.93	11.01	8.64	8.40	10.52	–	82.40	82.78	78.66
12	1.88	3.54	5.17	3.10	2.99	2.68	–	93.80	94.02	94.63
21	0.59	10.42	1.98	1.88	1.92	2.76	–	94.43	94.26	91.74
22	3.66	6.83	6.56	3.40	3.50	4.65	–	92.07	91.76	89.20
23	3.81	5.24	5.83	6.48	4.75	4.48	–	87.02	90.47	91.00
24	1.68	8.02	7.98	4.19	4.12	3.40	–	89.49	89.52	91.31
31	0.28	3.11	3.34	2.48	3.26	1.58	–	95.00	93.41	96.82
41	2.25	2.73	3.71	10.79	6.27	2.49	–	78.41	87.47	95.02
42	0.93	1.39	5.72	2.93	1.21	1.26	–	93.07	97.10	96.99
51	0.09	16.04	2.59	3.00	2.78	3.49	–	90.74	91.49	88.97
52	1.38	16.22	7.11	4.17	3.18	2.92	–	78.8	83.69	83.97
53	2.50	17.60	8.52	7.41	5.67	3.11	–	47.24	53.06	66.6
54	0.99	4.97	6.73	3.67	2.71	3.13	–	92.65	94.57	93.72
61	0.66	17.28	4.85	2.02	2.41	1.23	–	75.38	71.08	81.39
71	0.43	39.95	3.14	1.85	1.86	4.90	–	90.52	90.54	70.93
Mean	2.19	10.82	5.62	4.40	3.67	3.51	–	85.40	87.01	87.40
Median	1.38	8.02	5.72	3.40	3.18	3.11	–	90.52	90.54	91.00
Max	11.69	39.95	11.01	10.79	8.40	10.52	–	95.00	97.10	96.99
Min	0.09	1.39	1.98	1.85	1.21	1.23	–	47.24	53.06	66.60

The Symbol "–" Denotes that No Value is Available, and the Values of the Proposed Method are Highlighted.

9.3.2 Parameter Sensitivity Analysis

To evaluate the robustness of the algorithm's sensitivity to parameters, experiments were conducted with different combinations for seven parameters. W_{seed} values ranged from 20 m to 28 m with an interval of 2 m; W_{scale} values ranged from 5 m to 8 m with an interval of 1 m; h_1 values ranged from 0.5 m to 0.8 m with an interval of 0.1 m; h_2 values ranged from 0.2 m to 0.4 m with an interval of 0.1 m; h_3 values ranged from 0.2 m to 0.4 m with an interval of 0.1 m; W_{grid} values varied between 1.0 m and 1.5 m with an interval of 0.5 m; N_t was uniformly set to 10. Thus, each dataset had 1440 parameter combinations, resulting in 21,600 filtering results for 15 datasets. The results indicates that 92.50% of the total errors fall between 1.00% and 7.00%. Additionally, the average total error for 15 sample datasets under each parameter configuration was calculated and is displayed in Fig. 10. The average total errors range from 3.47% to 4.81%. These results demonstrate that the proposed method is not sensitive to parameter value configurations within a reasonable range, ensuring robust filtering performance.

Moreover, low point density significantly impacts point cloud segmentation, thus decreasing filtering quality. However, the generation of pseudo-grids could address this issue by interpolating some pseudo-grid points, thereby improving algorithm performance. To evaluate the effect of pseudo-grids, Sample 54 was used. Sample 54 is a point cloud dataset located in a rural area, with large point spacing ranging approximately from 2.0 to 3.5 m. The filtering result without the generation of pseudo-grids shows a total error of 13.08%. When pseudo-grids are generated with a grid size of 1.0 m, the total error decreases to 3.13%. With a grid size of 1.5 m, the total error increases to 5.35%. This comparison indicates that the proposed method enhances filtering quality through pseudo-grid generation, and it also highlights that increasing the grid size affects the quality of the filtering results.

9.3.3 Experiments on Large Regional Data

To further validate the performance of the algorithm, the Netherlands large-scale data were selected for experimentation. The Netherlands dataset covers an area of 11.25 km^2, with elevation ranging from 67.40 m to 145.20 m, and approximately 420 million point cloud data. The area includes features such as fractured terrain, steep slopes, dense low vegetation, rivers, and large buildings. Besides, the adaptive partition-based point cloud filtering method was used. Manual checking demonstrates that the algorithm performs well in complex areas with very few Type I error point clouds. However, compared to Type I errors, there are more inaccurately extracted ground point clouds, primarily due to errors in ground seed point extraction, and sub-segmentation, leading to Type II error points such as objects attached to the ground surface like bridges, buildings, and water bodies, as indicated by the black rectangles labeled A, B, and C. Additionally, during the final ground point cloud extraction process, some low vegetation point clouds are incorrectly extracted.

To evaluate the performance, Pingel's algorithm (http://tpingel.org/code/smrf/smrf.html) and commercial software Terrasolid TerraScan® are used for data processing, and the results are quantitatively compared with this algorithm, as shown in Table 9.3. From Table 9.3, it can be observed that both in terms of total error and kappa coefficient, this method outperforms the others, although it does not perform as well as TerraScan in Type II error.

TABLE 9.3
Performance Comparison Between the Proposed Method, the Method Described by Pingel et al. (2013), and the Commercial Software TerraScan®

The Proposed Method (%)				Pingel (2013) (%)				TerraScan			
T.I	T.II	T.E	k	T.I	T.II	T.E	k	T.I	T.II	T.E	k
0.37	2.82	1.62	96.76	2.07	3.76	2.93	94.14	8.97	0.81	4.83	90.33

Figure 9.5 shows the extraction process of breaklines and the generation of high-quality DEMs incorporating breaklines within the study area. To quantitatively evaluate the quality and positional accuracy of the extracted breaklines, three areas (P1, P2, P3) from Figure 9.5a were selected. Initially, reference breaklines were manually digitized on the DEMs, and the lengths of correctly extracted breaklines (TP), omitted breaklines (FN), and erroneous breaklines (FP) were measured through a comparison with the reference breaklines. Three metrics (Completeness, Correctness, and Quality) were computed following the methodology of Yang et al. (2013). Table 9.4 demonstrates that the proposed approach achieves a mean completeness of 90.21%, a mean correctness of 97.60%, and a mean quality of 88.26%, respectively. Furthermore, 2246 three-dimensional breakline points were manually extracted from 45 breaklines as reference points to evaluate positional accuracy. These points encompass jump, crease, and curvature breaklines. Subsequently, the reference points were matched with the nearest breakline points extracted by the proposed method, and the horizontal and vertical deviations were computed for each pair. The mean horizontal and vertical deviations were calculated as 0.23 m and 0.10 m, respectively. These findings indicate the proposed method's capability to extract breaklines with satisfactory quality and positional accuracy.

9.4 CONCLUSION

To address the ground filtering challenges in complex scenes and diverse terrains, as well as the inability of single primitives to robustly describe the differences between ground and non-ground point clouds, this chapter proposes an adaptive partition-based point cloud filtering method. Initially, to address issues such as low point cloud density and uneven distribution, the point cloud is processed through virtual gridization and divided into grid points and other point cloud categories. Subsequently, utilizing point cloud segmentation techniques, the grid point cloud is segmented, and, categorized into two types of regions. These regions are represented using surface element and point element primitives, and processed using point cloud segmentation filtering methods and an improved multi-scale morphological filtering method to obtain ground point clouds. Finally, the extracted ground points are utilized to generate a DEM as a ground reference model. Additionally, this method enables the extraction of multiple types of three-dimensional breaklines simultaneously from the primitives. To validate the algorithm's performance, experiments were conducted using the ISPRS test dataset and a large-scale dataset from domestic and international sources. Compared with other methods, experimental results demonstrate that

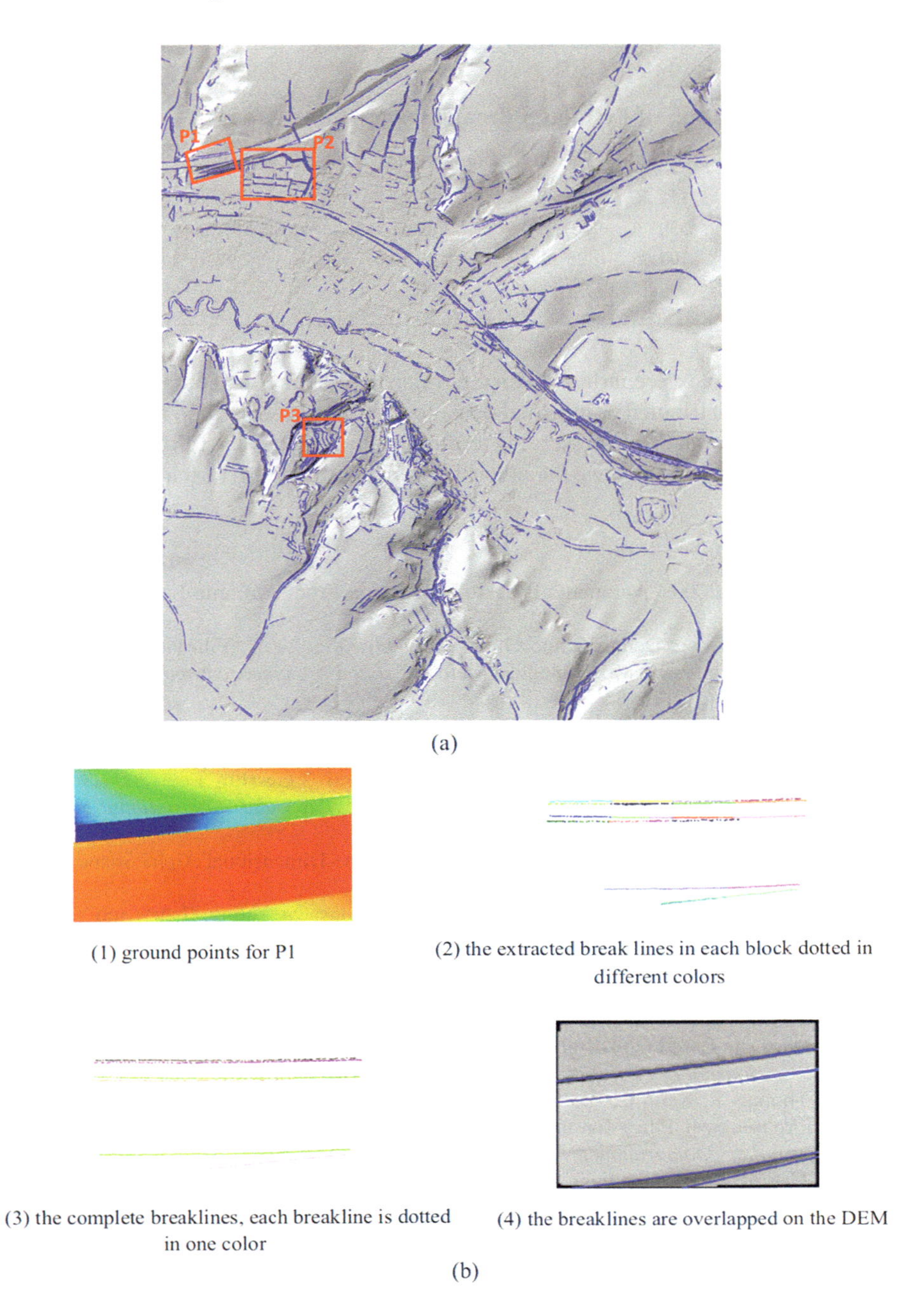

FIGURE 9.5 Extracting breaklines and generating DEMs with breaklines: (a) DEMs generated with breaklines, (b) the process of extracting the breaklines from ground points in area P1 (Yang et al. 2016).

TABLE 9.4
Quality Evaluation of Extracted Breaklines for three Areas

Area	TP (m)	FN (m)	FP (m)	Completeness (%)	Correctness (%)	Quality (%)
P1	1182.94	112.26	23.42	91.33	98.06	89.71
P2	4750.8	556.56	129.19	89.51	97.35	87.39
P3	946.61	107.65	25.23	89.79	97.4	87.69
Mean				90.21	97.6	88.26

in various complex scenarios, diverse terrains, and situations characterized by low and unevenly distributed point cloud densities, the proposed method robustly extracts ground point clouds while preserving terrain features effectively. Furthermore, within reasonable parameter ranges, the algorithm remains stable performance.

REFERENCES

Brügelmann, R. 2000. "Automatic Breakline Detection From Airborne Laser Range Data." *IAPRS.*

Chen, Qi, Peng Gong, Dennis Baldocchi, and Gengxin Xie. 2007. "Filtering Airborne Laser Scanning Data with Morphological Methods." *Photogrammetric Engineering & Remote Sensing* 73 (2): 175–85.

Chen, Q., 2009. "Improvement of the Edge-based Morphological (EM) method for lidar data filtering." *International Journal of Remote Sensing*, 30, 1069–1074.

Cho, Woosug, Yoon-Seok Jwa, Hwi-Jeong Chang, and Sung-Hun Lee. 2004. "Pseudo-Grid Based Building Extraction Using Airborne Lidar Data." *International Society of Photogrammetry and Remote Sensing* 35: 378–81.

Edelsbrunner, Herbert, and Ernst P. Mücke. 1994. "Three-Dimensional Alpha Shapes." *ACM Transactions on Graphics* 13: 4–72.

Hu, Han, Yulin Ding, Qing Zhu, Bo Wu, Hui Lin, Zhiqiang Du, Yeting Zhang, and Yunsheng Zhang. 2014. "An Adaptive Surface Filter for Airborne Laser Scanning Point Clouds by Means of Regularization and Bending Energy." *ISPRS Journal of Photogrammetry and Remote Sensing* 92: 98–111.

Mongus, Domen, Niko Luka, and Borut Alik. 2014. "Ground and Building Extraction From Lidar Data Based On Differential Morphological Profiles and Locally Fitted Surfaces." *ISPRS Journal of Photogrammetry and Remote Sensing* (7): 145–56.

Pingel, Thomas J., Keith C. Clarke, and William A. Mcbride. 2013. "An Improved Simple Morphological Filter for the Terrain Classification of Airborne Lidar Data." *ISPRS Journal of Photogrammetry & Remote Sensing* 77 (Mar.): 21–30.

Sithole, George, and George Vosselman. 2004. "Experimental Comparison of Filter Algorithms for Bare-Earth Extraction from Airborne Laser Scanning Point Clouds." *ISPRS Journal of Photogrammetry and Remote Sensing* 59 (1–2): 85–101.

Tóvári, Daniel, and Norbert Pfeifer. 2005. "Segmentation Based Robust Interpolation-a New Approach to Laser Data Filtering." *International Archives of Photogrammetry, Remote Sensing and Spatial Information Sciences* 36 (3/19): 79–84.

Yang, Bisheng, Lina Fang, and Jonathan Li. 2013. "Semi-Automated Extraction and Delineation of 3D Roads of Street Scene from Mobile Laser Scanning Point Clouds." *ISPRS Journal of Photogrammetry and Remote Sensing* 79: 80–93.

Yang, Bisheng, Ronggang Huang, Zhen Dong, Yufu Zang, and Jianping Li. 2016. "Two-Step Adaptive Extraction Method for Ground Points and Breaklines from Lidar Point Clouds." *ISPRS Journal of Photogrammetry and Remote Sensing* 119: 373–89.

10 3D Building Reconstruction

10.1 METHODOLOGY

A flowchart of the presented building reconstruction scheme is illustrated in Figure 10.1. Taking the building point cloud as the input, a pre-processing step, including roof extraction and segmentation, is adopted on the roof point clouds (a). A Roof Attribute Graph (RAG) is generated between the roof planes (b), and planar segments are grouped into a meaningful structure based on the RAG and an unambiguous top-down decomposition (c). We also adopt a bottom-up refinement to enhance the grouped structures and construct visually completed 3D models. The final output contains hierarchical roof structures with their associated topology (d). Gestalt laws for the reconstruction are proximity, similarity, continuity, and closure. Detailed information about the pre-processing steps on the roof extraction and segmentation can be found in (Yang et al. 2016).

10.1.1 Construction of RAG for Roof Hierarchical Structures

In this section, we first introduce the representation of building roofs using an RAG. Then, we describe the generation of an RAG and hierarchical structure grouping based on Gestalt principles.

10.1.2 Roof Representation and Gestalt Laws

Every complicated building can be reconstructed from a set of simple and meaningful structural models, grouped from various parameterized primitives based on an RAG. In the procedure, primitives are first grouped into sub-graphs, which are then combined into a hierarchical structure tree for the building roof. The tree root is the whole model, while the structures and extracted roof primitives can be taken as child nodes or leaves. A structure usually combines two or more parametric planar primitives or a continuous surface. In this chapter, we study the structured scenes, where planes are the main primitives and are considered as a vertex of the RAG.

To group the sub-graph (roof structures) and construct a more detailed and meaningful RAG, Gestalt laws (Kubovy and Berg 2008; Wang et al. 2015) are introduced, which summarize how humans perceive the form, model, and semantics of each part of a roof. The rules are described as follows:

Law 1. Proximity (F_d)—primitives that are closer together can be regarded as one group. We defined two planes as adjacent when the associated boundaries intersect as a line.

DOI: 10.1201/9781003486060-14

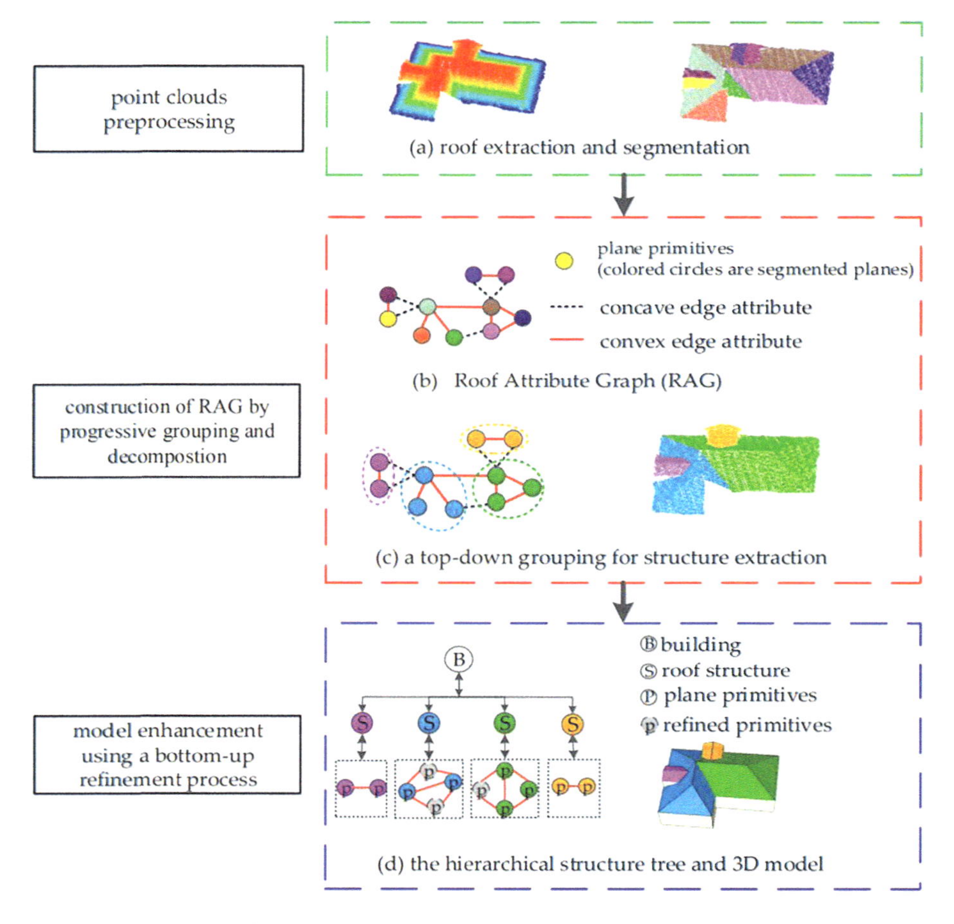

FIGURE 10.1 Overview of the method. (Hu et al. 2018).

Law 2. Similarity (F_{con})—primitives that share visual characteristics, such as shape, convexity, or concavity, form a perceptive group. The F_{con} is calculated by:

$$F_{con} = \begin{cases} Fconvex = \theta 1 < \theta 2 \\ Fconcave = \theta 1 < \theta 2 \end{cases} \tag{10.1}$$

Law 3. Continuity (F_{cco})—preference for continuous convexity/concavity, and shapes are aligned as one group. Figure 10.2 demonstrates how planes A and B are only labeled as one group if they satisfy one of two conditions. Planes A and B are linked with a convex edge in an RAG (a), and if a shared plane, S, exists between current planes or a neighboring group of planes (b). The connection attribute, F_{con}, between the two pairs must be the same.

Law 4. Closure (F_o)—if a shape is indicated, the whole group is obtained by filling in the missing data, thus, closing simple figures, and closure produces a meaningful and pleasant visualization of a structural model.

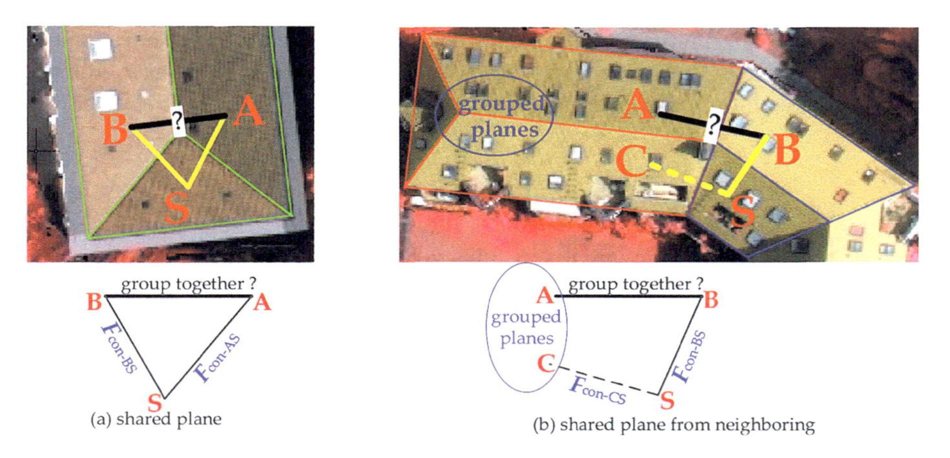

FIGURE 10.2 Illustration of the continuity law for grouping between planes (a, b). (Hu et al. 2018).

Following the application of these Gestalt laws, the segmented primitives from the pre-processing stage are used to construct the RAG, which contain no structural information. Thus, a hierarchical grouping strategy for roof structures was proposed based on the RAG.

10.1.3 Hierarchical Structure Grouping Using RAG

To find the meaningful structures from the RAG and generate a hierarchical tree of a building, a progressive iterative grouping algorithm is proposed.

The hierarchical tree is obtained iteratively. In each iteration, we decompose the RAG using the progressive grouping method and update the hierarchical tree with grouped structures, eventually reaching a stable state.

To find a plausible decomposition of roof structures from the RAG, a progressive grouping is used, which is aimed at searching and finding the best set of planar primitives that potentially belong to the same group (a structure).

The details for one iteration of the grouping are as follows:

1. a planar primitive, L_{P0}, that the maximum geometric area is started from, and the plane group, $G = \{L_{P0}\}$, is initialized.
2. a candidate plane set, S, is created where all planar primitives are connected (an edge linked in RAG) to the last added plane, P_e, from the plane group, G. If the candidate set is empty or all primitives in such a set are already grouped, then the current grouping loop is terminated;
3. the candidate primitives from S that not satisfy the convexity/concavity constraint, Fcon (**Law 2**), and consistency constraint, Fcc (**Law 3**) are removed;
4. the remaining candidate primitives based on the Euclidian distance (F_d) between the candidate plane and plane, P_e, are sorted, and the candidate with the minimum connecting distance into G is grouped. If there are no remaining primitives, we exit;
5. progress to step (2) and continue to find a roof structure.

After a group of primitives are found, we update the tree nodes and remove the vertexes in the RAG. Thus, we then proceed to the next iteration with a new grouping process until all primitives in the RAG are grouped.

10.1.4 Model Enhancement and Refinement

LiDAR data are usually incomplete due to the limitations of acquisition devices and occlusions. Therefore, the initial structures generated by the progressive decomposition and grouping process are usually ambiguous and could not be interpreted or correctly identified as meaningful structures.

The final step of building roof reconstruction is to refine and enhance the reconstructed model. Initial grouped structures refined structures and the associated graph nodes from the initial hierarchical tree are synchronously upgraded.

To accomplish these enhancements, we projecte the normal vector of adjacent primitives onto the ground plane, where the $\vec{n}_{p1}$ and $\vec{n}_{p2}$ are the normal vectors.

An analysis is performed on whether the projected normal vectors are mutually parallel with respect to their intersection. Hence, the symmetry/non-symmetry is calculated by:

$$F_{\text{symmetry}} = \begin{cases} \text{true } \vec{n}_{p1} . \vec{n}_{p2} = \pm 1 \\ \text{false otherwise} \end{cases} \quad (10.2)$$

Moreover, the indicator of full-symmetry is defined and calculated by:

$$F_{\text{Sym}-P_{12}} = \begin{cases} \text{full} - \text{symmetry Area}_{P_2 \leftarrow P_1} \geq \omega \,.\, \text{Area}_{P_1} \\ \text{partial} - \text{symmetry otherwise} \end{cases} \quad (10.3)$$

where the overlapped area threshold ω is empirically specified, depending on the data quality. According to the preceding assumption and structure knowledge, the process of model enhancement is achieved in an iterative manner. The key steps of the refinement operator, are elaborated as follows:

1. a node containing the grouped primitives is searched, and its corresponding sub-node (a child part) and inlier leaf nodes (planar primitives) are extracted;
2. the symmetry indicators of the neighboring structures are calculated, and symmetry evaluation processing is performed;
3. closed hull loops are detected, and the projected primitives are stitched together in sequence, based on the closure perception laws (Law 4). In addition, an add and union primitive operation is carried out; and
4. a similar regular process as in (Yang et al. 2016) is applied to the refined structure, and the parameters for the corresponding nodes and primitives in the hierarchical tree are automatically updated.

The iterative processing of structure enhancement is terminated until no structures could be reconstructed from the initial tree.

TABLE 10.1
Description of the Two Urban Area Datasets

Site	Guangdong (China)	Vaihingen (Germany)
Acquired Date	May 2016	August 2008
Acquisition System	Trimble Harrier 68i	Leica ALS50
Fly Height	800 m	500m
Buildings	83	56
Covered Area	340 × 360 m^2	150 × 220 m^2
Point Density	12–15 points/m^2	4–6 points/m^2

10.2 EXPERIMENTS

We implemented the proposed algorithms and mainly tested them on two Airborne Laser Scanning (ALS) datasets that differ in point density and urban characteristics. Various internal consistency metrics were used to evaluate the reconstructed 3D building models. The results were compared with other state-of-the-art studies to ascertain the effectiveness of the proposed method.

10.2.1 Datasets Description

The proposed approach was tested on two urban area datasets. The details of these urban datasets are listed in Table 10.1. We collected the Guangdong data, while the Vaihingen dataset is from the benchmark data of the "ISPRS Test Project on Urban Classification and 3D Building Reconstruction" (Rottensteiner et al. 2014)

10.2.2 Building Reconstruction Results

Roof plane segmentation was achieved by the method referred to in (Yang et al. 2016). In addition, the overlapped area threshold for symmetry processing was set to 0.7 for the Guangdong dataset and 0.95 for the Vaihingen dataset. The results from the Guangdong and Vaihingen datasets are illustrated in Figure 10.3. Each color in these figures represents an individual plane.

Figure 10.3 shows a detailed visual inspection of the reconstructed building model, where the extracted roof planes are shown in various colors. A building model is reconstructed by combining the detected roof planes into meaningful roof structures.

10.2.3 Analysis and Discussion

10.2.3.1 Comparison

The proposed approach can more effectively identify the unambiguous roof parts using only simple grouping rules as compared to the methods based on the commonly used or improved RTGs (Xiong, Elberink, and Vosselman 2014; Wang et al. 2015; Xiong et al. 2015; Xu, Jiang, and Li 2017).

FIGURE 10.3 Results of the plane segmentation and 3D reconstruction (Vaihinge, Area 3) (Hu et al. 2018).

It can be seen that different reconstructed models of the same building were used to illustrate the effectiveness of the proposed method, which produced more unambiguous and meaningful roof parts. The generated model of the method proposed by (Verma, Kumar, and Hsu 2006) was based on a fixed model library combining simple roof primitives, which reduces the flexibility of the reconstruction process as it requires an exhaustive search to extract previous complex building elements. The RTG-based approach (Xiong et al. 2014; Elberink and Vosselman 2009; Xiong et al. 2015) is the most popular means to search and match the roof target graph, which results in redundancy as the roof planes and the associated intersecting lines may match multiple targets. In addition, the improved RTG approach development by (Xiong et al. 2014; Xiong et al. 2015) adopts corners as the basic unit for building reconstruction, which is flexible, but the topologies between corners and linked lines are difficult to interpret. Recent improvements by (Jarząbek-Rychard & Borkowski 2016) on this method can generate a topological unit without independent overlapping planes, but it is still hindered by the limited model library and might also identify the same RTG as different objects. Our proposed method, however, can avoid this ambiguous matching issue during reconstruction processing.

10.2.3.2 Evaluation on Vaihingen Dataset

These 3D models reconstructed from the Vaihingen dataset were evaluated by the International Society for Photogrammetry and Remote Sensing in terms of their topology correctness and model precision.

10.2.3.2.1 Topology Analysis

Figure 10.4 demonstrates the topological consistency evaluation results of the Vaihingen dataset by ISPRS. Based on the extraction results, the yellow label indicates that the pixels inside reference planes had a (1:1) relation to planes. The blue label means there was no corresponding plane and the dark magenta represents pixels in reference roof planes that were merged with other planes in the reconstruction results (1:N relation). The bright magenta label means that the pixels in the reference roof planes split into two or more planes during reconstruction (M:1 relation), while a dark cyan label indicates that the pixels in the roof planes were part of a plane cluster having (M:N) relations between planes in the reference image and planes in the reconstruction results. The relationship (M:1), suggests oversegmentation issues, indicating that a reconstructed plane refers to more than two planes in the reference dataset, and the (1:N) relation represents undersegmentation.

The building reconstruction assessment results show a precise overlap (1:1) for 80 roof instances out of 133 (approximately 60.15%). The reason for the effective roof plane extraction was that the building roofs were extracted from a coarser level to a finer level using a morphological scale space. The blue colors in the figure indicate that planes were either fully modeled or nearly completely neglected, a scenario caused by the missed point cloud during the building extraction or no LiDAR data. We also compared the reconstruction approach with four state-of-the-art algorithms from the website of ISPRS, with the details listed in Table 10.2.

It can be seen from the table that there was only one plane with (N:M) relations, where both over- and undersegmentation occurred, and the primary reason for the low errors was the symmetry calculation after roof extraction. The most common

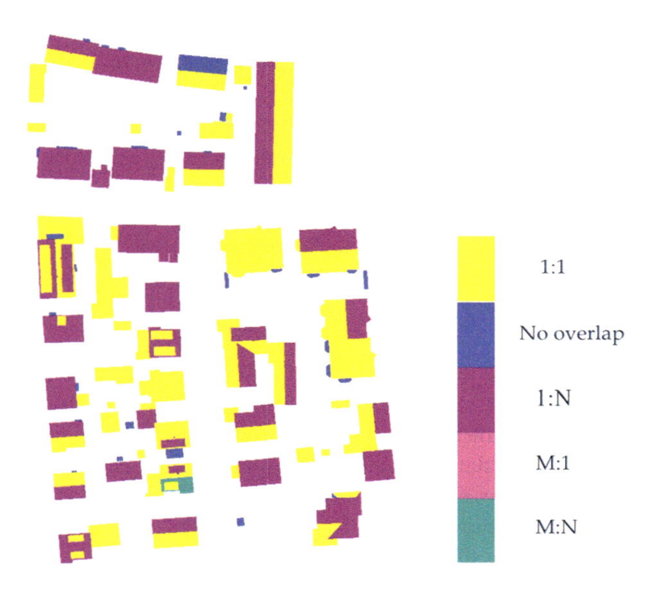

FIGURE 10.4 Topological differences between the 3D roof planes (results: Reference). (Hu et al. 2018).

TABLE 10.2
Comparison of Topological Errors Between the Reference and Reconstructed (Reference: Results)

Region	Method	(1:M) Oversegmentation	(N:1) Undersegmentation	(N:M) Over- and Undersegmentation
Vaihingen Area 3	CKU	4	48	2
	ITCX3	3	50	2
	YOR	2	51	1
	WROC_2b	3	52	3
	Proposed	0	48	1

difference was the (N:1) relation (undersegmentation), with a total number of 48 reconstructed instances. The primary reason for this difference was caused by the lost small structures overlapped in the base roof planes, but the final reconstructed models have not been affected according to the assessment. More importantly, the topological consistency in the relationship, (1:M), related to oversegmentation errors was zero, which indicates that there were no reconstructed roof planes resulting from a split plane. These successes can be attributed to roof segmentation in a morphological scale space.

10.2.3.2.2 Model Precision

The evaluation of the dependency between the plane size and model quality. It can be seen from the individual histogram that there was a leap between the completeness and quality (planes with an area of 5m^2). This is due to symmetry processing during model enhancement. In addition, the reconstruction completeness was greatly influenced by the accuracy of the roof extraction, as complex roof structures were more likely to be organized by small patches than large planes. These conclusions are in line with the methods based on an RTG library. (Xiong, Elberink, and Vosselman 2014; Jarząbek-Rychard & Borkowski 2016)

Comparative results were evaluated on a per-roof plane level in two ways: All extracted roof planes and large roof planes that covered an area of more than 10m^2. Four state-of-the-art methods described on the website of ISPRS were selected for comparison. The measures for the completeness and correctness of per-objects are displayed in Table 10.3.

As seen in the table, the differences in the metrics between all extracted and large roof planes indicate that completeness performance was related to the plane size. In addition, about 14.5% of the roof planes were not reconstructed. The main reason for this was the missing input data, caused by the building point cloud detection, as small patches had too few points for extraction, or the small roofs were neglected during regularization. A reasonable approach to overcome this shortcoming is to improve the pre-processing step for the extraction of building points.

For most of these methods, one of the two indicators exceeded the median value, while the other indicator was reduced. We obtained an 85.5% completeness and 97.7% correctness with our proposed method. The values of the two metrics

TABLE 10.3
Comparison of the Final Reconstruction Performance (Per-roof plane)

		All Roof Planes		Large Roofs with Areas Greater than 10 m^2	
Region	**Method**	**Completeness**	**Correctness**	**Completeness**	**Correctness**
Vaihingen Area 3	CKU	81.3%	98.4%	91.9%	99.1%
	ITCX3	88.1%	88.2%	96.8%	95.8%
	YOR	84.7%	100.0%	97.6%	100.0%
	WROC_2b	81.7%	100.0%	92.7%	100.0%
	Proposed	85.5%	97.7%	96.0%	100.0%

approached the median, which means that we achieved a relative balance between completeness and correctness. The reason for this balance was the roof plane segmentation method (Yang et al. 2016), but more importantly, an unambiguous principal direction for the regularization was easily obtained from the locally clustered building structure. The average root mean squared error (RMSE) values were 0.5 m and 0.29 m in the horizontal and vertical directions, respectively, which indicates that the proposed approach has a high geometric accuracy, as compared with the four state-of-the-art methods.

10.3 CONCLUSION

We present a novel method for roof modeling from 3D point clouds based on the Roof Attribute Graph (RAG). The RAG is a logical model for building roofs, while the grouped sub-graph is a structure that incorporates geometric parameters. The output of the reconstruction process is a complete polyhedral structured model, and mutual relationships are organized by a hierarchical topology tree. By first segmenting the roof planes and constructing an RAG, we develop category-specific reconstruction methods to obtain visually pleasing structural models, even in the presence of occlusions and incomplete data. The key idea is to extract potential structures by hierarchical grouping, while an analysis-and-enhancement scheme is developed to improve the basic structure best fitting to both human vision and knowledge constraints.

Our structural modeling scheme has a few limitations that lead to failures in the final model, including missing neighboring segments and sparse points, which resulted in ambiguous structural enhancement and segmentation errors for non-planar objects. The former two scenarios can be avoided by using a higher density and quality of points, while a new strategy for non-plane objects needs to be developed, as in (Dong et al. 2018). We believe that there are many opportunities for further exploration. For example, we should be able to add more data, collected by mobile or ground laser scanners, into the RAG to construct a more reasonable and feasible building structural model to support advanced editing and calculation at LoD2 or LoD3, and images can be used to extract fine details and unambiguous semantic information during building reconstruction. Overall, these structural models with unambiguous geometries and topological relationships are expected to lead to better models that support not only high-fidelity visualization but also editing, and could eventually become searchable.

REFERENCES

Dong, Zhen, Bisheng Yang, Peijun Hu, and Sebastian Scherer. 2018. "An Efficient Global Energy Optimization Approach for Robust 3D Plane Segmentation of Point Clouds." *ISPRS Journal of Photogrammetry and Remote Sensing* 137: 112–33. doi:10.1016/j.isprsjprs.2018.01.004

Elberink, Sander Oude, and George Vosselman. 2009. "Building Reconstruction by Target Based Graph Matching on Incomplete Laser Data: Analysis and Limitations." *Sensors* 9 (8): 6101–118. doi:10.3390/s90806101

Hu, Pingbo, Sheng Yang, Xiangqian Wu, Lingli Zhu, and Jizhou Sun. 2018. "Towards Reconstructing 3D Buildings from ALS Data Based on Gestalt Laws." *Remote Sensing* 10 (7): 1127. doi:10.3390/rs10071127

Jarząbek-Rychard, M., & Borkowski, A. 2016. "3D building reconstruction from ALS data using unambiguous decomposition into elementary structures." *ISPRS Journal of Photogrammetry and Remote Sensing*, 118, 1–12. doi: 10.1016/j.isprsjprs.2016.04.005

Kubovy, Michael, and Michał van den Berg. 2008. "The Whole Is Equal to the Sum of Its Parts: A Probabilistic Model of Grouping by Proximity and Similarity in Regular Patterns." *Psychological Review* 115 (1): 131–54. doi:10.1037/0033-295X.115.1.131

Rottensteiner, Franz, Gunho Sohn, Markus Gerke, Jan Wegner, Uwe Breitkopf, and Jaewook Jung. 2014. "Results of the ISPRS Benchmark on Urban Object Detection and 3D Building Reconstruction." *ISPRS Journal of Photogrammetry and Remote Sensing* 93: 256–71. doi:10.1016/j.isprsjprs.2013.10.004

Verma, Vijay, Rajiv Kumar, and Shin Hsu. 2006. "3D Building Detection and Modeling from Aerial LiDAR Data." In *Proceedings of the 2006 IEEE Computer Society Conference on Computer Vision and Pattern Recognition*:2213–20. doi:10.1109/CVPR.2006.12

Wang, Hongtao, Wuming Zhang, Yiming Chen, Mei Chen, and Kai Yan. 2015. "Semantic Decomposition and Reconstruction of Compound Buildings with Symmetric Roofs from LiDAR Data and Aerial Imagery" *Remote Sensing* 7: 13945–74. doi:10.3390/rs71013945

Xiong, Bo, Sander Oude Elberink, and George Vosselman. 2014. "A Graph Edit Dictionary for Correcting Errors in Roof Topology Graphs Reconstructed from Point Clouds." *ISPRS Journal of Photogrammetry and Remote Sensing* 93: 227–42. doi:10.1016/j.isprsjprs.2014.01.007

Xiong, Bo, Michal Jancosek, Sander Oude Elberink, and George Vosselman. 2015. "Flexible Building Primitives for 3D Building Modeling." *ISPRS Journal of Photogrammetry and Remote Sensing* 101: 275–90. doi:10.1016/j.isprsjprs.2015.01.002

Xu, Bo, Wanshou Jiang, and Lelin Li. 2017. "HRTT: A Hierarchical Roof Topology Structure for Robust Building Roof Reconstruction from Point Clouds." *Remote Sensing* 9: 354. doi:10.3390/rs9040354

Yang, Bisheng, Ronggang Huang, Jianping Li, Mao Tian, Wenxia Dai and Ruofei Zhong. 2016. "Automated Reconstruction of Building LoDs from Airborne LiDAR Point Clouds Using an Improved Morphological Scale Space." *Remote Sensing* 9: 14. doi:10.3390/rs9010014

11 3D Road Reconstruction

11.1 METHODOLOGY

11.1.1 ROAD BOUNDARY EXTRACTION AND VECTORIZATION

11.1.1.1 Methodology Overview

The proposed method mainly consists of two steps: road boundary extraction and accurate 3D road boundary vectorization. Figure 11.1 illustrates the workflow of the proposed method. A pre-processing component is first performed to locate the approximate road areas for discarding the off-ground points according to the relative elevation. Then, the curb points are obtained with a supervoxel-based method in terms of fine border preservation. The candidate curb supervoxels are further clustered to locate the road boundary by a contracted distance clustering strategy. Finally, the extracted 3D road boundary is represented through a series of cubic Bezier curves following a Kalman filter to complete the missing parts. The vectorized 3D road boundaries are used to calculate road-related parameters such as boundary locations, road widths, turning radiuses, and slopes, for road survey and analysis.

11.1.1.2 Road Boundary Extraction

11.1.1.2.1 Supervoxel-Generation

We propose an efficient supervoxel-generation method for road boundary extraction. Before the supervoxel-generation, the ground points are divided into 2D grids with a fixed grid size, as shown in Figure 11.2. The elevation difference of each grid is calculated by Eq. (11.1). The grids with elevation differences larger than the predefined threshold H_T are labeled as uneven grids (UG). The other grids are labeled as smooth grids (SG).

$$h_i^{\text{range}} = h_i^{\max} - h_i^{\min} \tag{11.1}$$

Where $h_i^{\max}$ and $h_i^{\min}$ are the maximum and minimum elevation in the grid i, respectively; h_i^{range} is the height difference in grid i.

Because only the uneven grids may contain the points that belong to the curbs. the points located in UG are resegmented into supervoxels with better-detailed border preservation, which will ultimately benefit the later road curb extraction. The supervoxels are generated according to the weighted distances. The weighted distances between each candidate point and its neighboring SG and UG, are calculated based on point density differences, elevation differences, and horizontal distances by Eq. (11.2) and (11.3), respectively. The reason for ignorance of the elevation difference is that the curb surface is not a horizontal flat, and points on it never share identical elevation values. The point densities of the candidate points are calculated by Eq. (11.4). The point densities of the uneven grids are calculated as the largest point

DOI: 10.1201/9781003486060-15

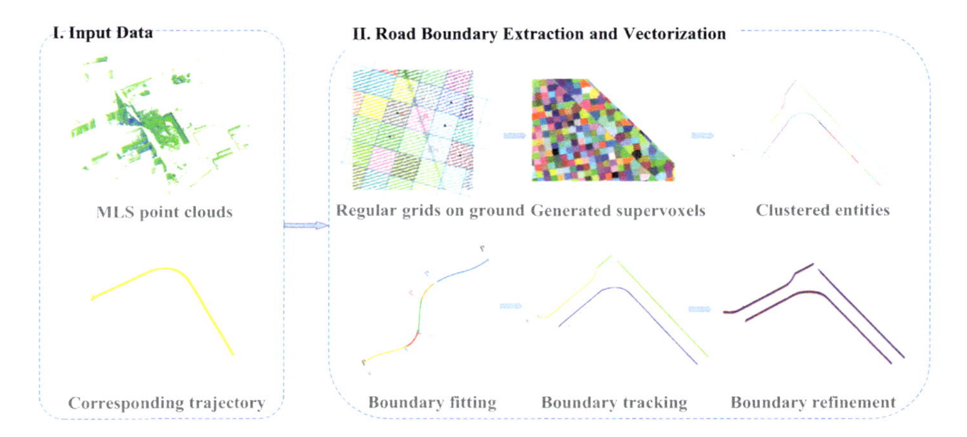

FIGURE 11.1 Workflow of the proposed road boundary extraction and vectorization method.

FIGURE 11.2 Illustration of supervoxel-generation (Mi et al. 2022).

density of all points belonging to it, while the point density of the smooth grids is calculated by Eq. (11.5). Finally, the candidate points are reassigned to the uneven or smooth grid according to the minimum weighted distance. Figure 11.2 illustrates an example of the supervoxel-generation for curb extraction.

$$D(p, \mathrm{SG}) = w_{xy}D_{xy} + w_z D_z + w_{ds}D_{ds} \tag{11.2}$$

$$D(p, \mathrm{UG}) = (w_{xy} + w_z)D_{xy} + w_{ds}D_{ds} \tag{11.3}$$

where p is the candidate point in uneven grid; $D(p, \mathrm{SG})$ is the characteristic difference between p and its neighboring smooth grid, so is $D(p, \mathrm{UG})$; D_{xy}, D_z, and D_{ds} are the differences of planimetric Euclidean distances, elevation differences, and local density differences between p and SG(UG), respectively; w_{xy}, w_z, and w_{ds} are the weights of the D_{xy}, D_z, and D_{ds}, respectively.

$$p_{ds} = \frac{N_r}{\pi r^2} \tag{11.4}$$

$$\mathrm{SG}_{ds} = \frac{N}{A(\mathrm{SG}_{xy})} \tag{11.5}$$

where r is the searching radius for local density calculation of p, and N_r is the number of searched points within radius r; $A(SG_{xy})$ is the planimetric area of SG, and N is the number of points located in SG.

The supervoxel-generation method is specifically designed for road curbs in point clouds. Only points in UG are taken into account in the process of restructuring the inclusion relationship. And time-consuming calculation (like normal estimation) is not employed at all. Hence, the supervoxel-generation method has good computation efficiency.

11.1.1.2.2 Approximate Road Boundary Extraction

The approximate road boundaries are detected by clustering the generated uneven supervoxels into long and narrow segments based on a modified Euclidean distance clustering algorithm. Because most of the road boundary points are located in the uneven supervoxels after supervoxel-generation, as shown in Figure 11.2. However, it is worth noting that the point densities along the same scanning line are significantly larger than those between adjacent scanning lines caused by the MLS mechanical measurement. Therefore, the classical Euclidean distance clustering algorithm may produce unsatisfactory clusters as it is difficult to specify the cluster searching radius. Thus, a distance-contracted clustering method is adopted to overcome the influence of varying point densities. The contracted distance between the points p_i and p_j is calculated by Eq. (11.6). The contracted distance is used to cluster the initial road boundary points, as shown in Figure 11.3.

$$d_c\left(p_i, p_j\right) = d\left(p_i, p_j\right) * \left(1 - cos\left\langle \overrightarrow{p_i p_j}, \overrightarrow{p_i^{traj}} \right\rangle\right)^2 \tag{11.6}$$

where p_i and p_j are the two points in adjacent uneven supervoxels, and $d(p_i, p_j)$ is the Euclidean distance between p_i and p_j; p_i^{traj} is the nearest trajectory point to p_i; $\overrightarrow{p_i^{traj}}$ denotes the heading of p_i^{traj}; $d_c(p_i, p_j)$ is the distance after contraction between p_i and p_j.

11.1.1.3 Road Boundary Vectorization

The approximate road boundary points are discrete point sets located on the road curbs, which are modeled as 3D vectorization representations for further purposes.

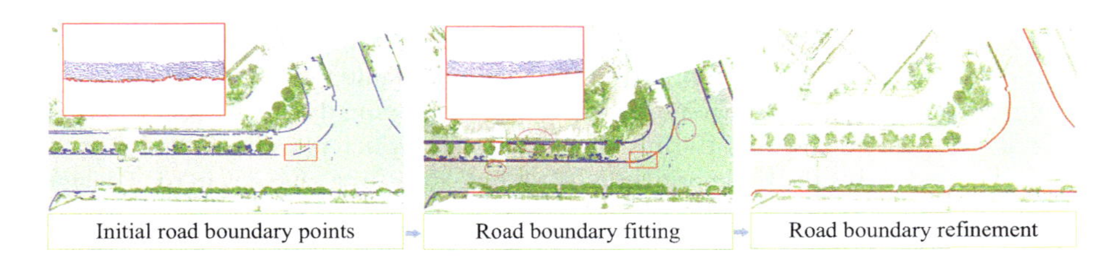

FIGURE 11.3 Illustration of road boundary vectorization (Mi et al. 2022).

11.1.1.3.1 Road Boundary Fitting

As the curvature of the road boundary may change in a scenario, a lower-order polynomial model is difficult to fit well, while a higher-order polynomial model suffers from high computational complexity and poor stability. Therefore, a more flexible piecewise fitting model consisting of a chain of continuous cubic Bezier curves (Forrest 1980) is adopted. The flexible piecewise fitting model aims to achieve global optimization by fitting multiple high-precision segments. The number of curves is determined according to the complexity of the road boundary shapes. Boundaries with complex shapes will be divided into more segments using the DP (Douglas-Peucker) algorithm (Hershberger and Snoeyink 1992) and vice versa. Inspired by the process of detecting the preserved vertices in the DP algorithm, we employ the strategy to determine the segmentation points where the tangent changes drastically. Based on the segmentation, the tangent of the boundary in the same segment changes gently and is more inclined to fit well using a cubic Bezier curve, which is written by Eq. (11.7).

$$B_k(t) = (1-t)^3 CP_0 + t*(t-1)^2 CP_1 + t^2(t-1)CP_2 + t^3 CP_3, t \in [0,1] \tag{11.7}$$

$$t_j = \frac{\sum_{i=1}^{j} d(p_i, p_{i-1})}{\sum_{i=1}^{N} d(p_i, p_{i-1})} \tag{11.8}$$

where $B_k(t)$ is a set of ordered points used to fit a cubic Bezier curve for the k-th segment of an initial road boundary; CP_0, CP_1, CP_2, and CP_3 are the four calculated control points of the fitted cubic Bezier curve. For the j-th point in the ordered points set, $B_k(t_j)$ is its coordinates while t_j is calculated as Eq. (11.8), where $d(p_i, p_{i-1})$ is the Euclidean distance between p_i and p_{i-1}.

The superiority of the cubic Bezier curve is as follows: (1) good balance between computing efficiencies and fitting accuracies; (2) parametric representation for easy interpolation in mapping; and (3) the light calculation of tangents at endpoints and beneficial for road boundary tracking. Figure 11.3 shows the selected points on the bottom of the approximate road boundary and the fitted 3D road boundary in the enlarged image.

As a lot of road boundaries are missed because of occlusions and entrances, those extracted parts are tracked by a Kalman filter using the topology between road boundaries. And missing parts would be completed.

11.1.1.3.2 Road Boundary Tracking

The fitted road boundaries are firstly divided and sorted into the left boundary sets B_l and right boundary sets B_r, according to the vehicle moving direction. The curvature of road boundary changes uniformly in a local range, which is well justified in most scenarios and meets the basic principles of smooth curvature changes in road design (Holgado-Barco et al. 2015). Hence, a Kalman filter is adopted to track the detected road boundary to complete the road boundary with high continuity and integrity. The process for the Kalman filter includes the prediction processes and correction processes. The prediction process is to predict the potential boundary location from the previous information while the correction process is to verify the existence of the boundary at the predicted location (Welch and Bishop 1995). The equation in the state prediction process is written by Eq. (11.9).

$$X_k^{pred} = AX_{k-1} + W_k \quad (11.9)$$

$$A = \begin{bmatrix} I_{3*3} & diag(s,s,s) & 0 \\ 0 & I_{3*3} & diag(s,s,s) \\ 0 & 0 & I_{3*3} \end{bmatrix} \quad (11.10)$$

where X_k^{pred} is a vector representing the predicted state at the forward step k; X is a 9-dimensional vector $\left[x, y, z, t_x, t_y, t_z, t_x', t_y', t_z'\right]$, where $[x, y, z]$ represent the coordinates of the end vertexes of the curb entity, $[t_x, t_y, t_z]$ represents its unit tangent, $\left[t_x', t_y', t_z'\right]$ is the corresponding first order derivative of the unit tangent; W_{k-1} is a process noise set as zero mean Gaussian white noise; I_{3*3} is an identity matrix in size of $3 * 3$; s is the forward step size for each state prediction.

$$Z_k^{pred} = HX_k^{pred} + V_k \quad (11.11)$$

where Z_k^{pred} is the vector of measurements as $[x, y, z, t_x, t_y, t_z]$; V_k is the measurement noise terming for each observation in the measurement vector and is assumed to be zero mean white Gaussian noise; H is the transformation matrix mapping state vector into measurement domain as $[I_{6*6}, 0_{6*3}]$.

$$\widehat{X_k} = X_k^{pred}(tracked) \quad (11.12)$$

For the correction process, if the end vertex of any existing road boundary entity is searched in the buffer zone of Z_k^{pred}, the searched road boundary will be accepted as the tracked boundary and the estimated state will be renewed as the state of the tracked boundary by Eq. (11.12). Otherwise, the estimated state is accepted as the predicted one $\widehat{X_k} = X_k^{pred}$.

11.1.1.3.3 Road Boundary Refinement

To remove false road boundaries caused by moving vehicles, pedestrians, pole-like objects, and walls. As for these false positives, not only is the length after tracking short, but the points elevation distribution of these objects is also significantly different from the true road boundaries. The on-ground and off-ground points are united to determine the continuous elevation histogram to distinguish the footprint of the wall from road boundaries (Li et al. 2019). Special emphasis is on the condition when the wall is the road boundary, which may appear in tunnels. In response to this situation, the footprint of the wall is considered as road boundary when it is the nearest detected boundary to the trajectory. A mutual overlap matrix is also employed to refine the extracted boundaries on the left and right sides separately. The false road boundaries are removed by Eq. (11.13).

$$L(G)=\alpha\left(1-\frac{\bigcup_{i=1}^{N}G(i,i)}{Traj}\right)+\sum_{i=1}^{N}\sum_{j=1,j\neq i}^{N}G(i,j) \tag{11.13}$$

$$G(i,i)=T(A_i) \tag{11.14}$$

$$G(i,j)=\left(T(A_i)\cap T(A_j)\right)/T(A_i) \tag{11.15}$$

where $G(i, i)$ is the corresponding trajectory interval of the i-th boundary A_i, and $G(i, j)$ is the coincidence ratio of the corresponding trajectory interval between A_i and A_j; α is the weighting parameter providing a trade-off between completeness and correctness.

11.1.2 Road Markings Extraction and Reconstruction

11.1.2.1 Methodology Overview

Figure 11.4 illustrates the workflow of the proposed two-stage approach for road marking extraction and modeling. Firstly, candidate road marking bounding boxes with semantic labels are detected on the feature map by a general object detection network, YOLOv3 (Redmon and Farhadi 2018). Then, the raw point cloud coordinate system determines fine positions, orientations, and scales of the detected road markings by optimizing an energy function from point clouds. Finally, a reranking strategy with optimal selection enhances performance by combining the detection confidence and the localization score to produce the final road marking models.

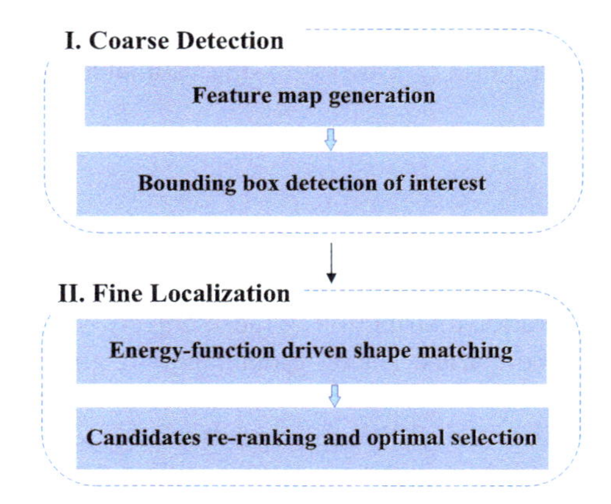

FIGURE 11.4 Workflow of the proposed road marking extraction and modeling method.

11.1.2.2 Candidate Road Marking Detection

To overcome the limitations of the bottom-up road marking extraction methods, candidate road marking bounding boxes with semantic labels are first detected on feature maps, as shown in Figure 11.5. Then these bounding boxes are back-projected from the local feature map coordinate system to the raw point cloud coordinate system. Figure 11.5 illustrates the candidate road marking detection procedure.

11.1.2.2.1 Feature Map Generation

Sectioning MLS point clouds into blocks along the vehicle trajectory will result in faster processing. As in (Rastiveis et al. 2020), the size of blocks varies according to roadway geometry. For straight sections, the length of a tile is fixed as 50m. For curved sections, smaller blocks are segmented according to a curve radius function. The constant maximum block width is 25m for all blocks to cover the entire road surface as much as possible. Overlapping of two adjacent blocks is also necessary to prevent gaps and ensure object completeness. The overlap distance is specified based on the maximum size of a road marking instance of interest, which is 9m in this research. Points in blocks above the trajectory were discarded (Mi et al. 2022), and

FIGURE 11.5 Candidate road marking detection (Mi et al. 2021).

the remaining points were projected on the geo-referenced feature maps in the bird's view according to the following steps.

Feature map rotation. The block is rotated around the z-axis to align the longitudinal road direction to the height of the feature map, which benefits to regress the minimum size of the bounding box.

Feature map resolution. The size of the road markings varies greatly. Moreover, road markings are slender in the longitudinal direction. For example, the length of a dashed lane boundary is several times greater than its width. However, it is challenging to achieve detection performance for slim objects equal to that of shapes with similar aspect ratios in the object detection task (Liu et al. 2020). For better performance, the width resolution of the feature map should be greater than the height resolution.

Feature map features. The average intensities, elevation variances, and distances from the points to the scanner center of the points in each pixel were used to generate the feature maps. Intensity is the most distinctive feature that distinguishes road markings from their surroundings. Because vehicles and dashed lanes look like thin bars in the feature map, elevation differences were used to distinguish them. Point cloud intensity attenuates with increasing distance, and therefore distances were chosen to compensate implicitly for this attenuation. Specifically, feature values were set to zero where LiDAR resolution was lower than that of the feature map or in occluded areas.

Figure 11.5 illustrates an example of feature map generation. To visualize the three input features in feature maps, they were associated with RGB channels, where the intensity was in the B-channel, elevation variance in the G-channel, and range from the pixel to the scanner center in the R-channel.

11.1.2.3 Road Marking Detection

A general object detection network can be used to predict road marking bounding boxes with semantic labels on the generated feature maps. YOLO-v3 (Redmon and Farhadi 2018) was used in this research due to its good implementation, portability, and performance. YOLOv3 is a typical one-stage object detection network that uses the hybrid residual network, called Darknet-53, together with feature pyramid networks (FPN) (Lin et al. 2017) to extract multi-scale features. The extracted feature maps at three different scales were then used to predict the bounding boxes. The tensor dimension of the predicted box was 17, including four bounding box parameters, one objectiveness prediction, and 12 class predictions. The objectiveness score indicates the overlaps between the predicted bounding boxes and the ground truth. Twelve types of road markings of interest were predicted in this study, including seven types of arrows (straight arrow, left turn arrow, right turn arrow, straight & left arrow, straight & right arrow, left & round arrow, and U-turn arrow), forbidden, diamond, stop line, crosswalk, and dashed lane, as shown in Figure 11.6.

The predicted bounding boxes with semantic labels from all feature maps were back-projected to the raw point clouds and collected as initial candidate road markings. The overlapping between the adjacent blocks ensures the completeness of the

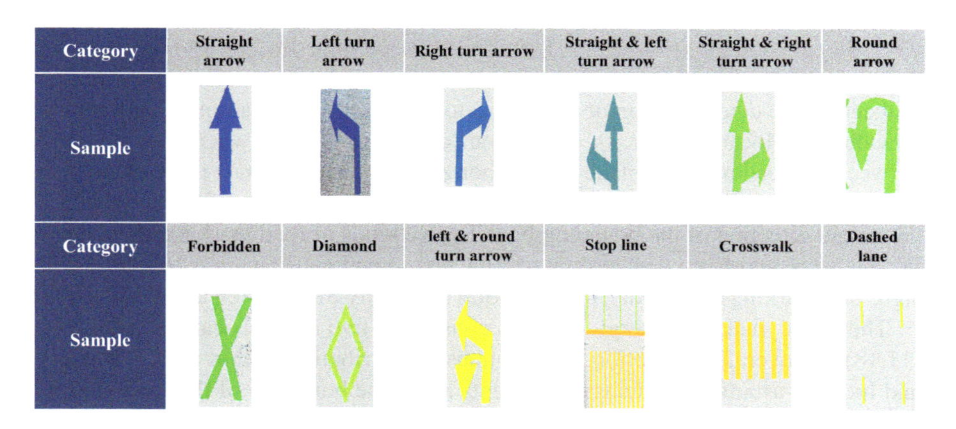

FIGURE 11.6 Twelve types of road markings of interest (Mi et al. 2021).

road markings that cross the cutting line. However, it also causes repeated detections in the overlapping area. For the sake of subsequent computation efficiency, non-maximum candidate bounding boxes were dropped when the overlapping ratio of the two candidates with identical semantic labels was greater than a threshold τ. Figure 11.5 illustrates global intraclass non-maximum suppression (NMS), where bounding boxes before and after the global intraclass NMS are shown at the bottom of Figure 11.5, respectively. The overlapping ratio was calculated by Eq. (11.16).

$$r_{\text{overlap}}^{P} = \frac{A(P \cap Q)}{A(P)} \tag{11.16}$$

where P and Q are two bounding boxes, $A(*)$ is the area, r_{overlap}^{P} is the overlapping ratio of P.

The suppression procedure was carried out using:

$$E_{\text{intra}} = E_{\text{coarse}} + \varepsilon \frac{b_w}{\max(b_w, s_w)} + (1-\varepsilon)\frac{b_h}{\max(b_h, s_h)} \tag{11.17}$$

where E_{coarse} is the confidence of the detected bounding box calculated by the object detection network; (b_w, b_h) are the width and height of the detected bounding box; (s_w, s_h) are the corresponding standard width and height referring to Chinese national standards for road traffic signs and markings (General Administration of Quality Supervision and China 2009); and ε is a weight ranging from 0 to 1.

11.1.2.4 Fine Localization and Modeling

To obtain the fine locations, orientations, and scales of road markings in the raw point cloud coordinate system, an energy function-driven shape matching and candidate reranking procedure with optimal selection were implemented. Figure 11.7 illustrates the procedure of the fine localization and modeling procedure for road markings.

11.1.2.4.1 Energy Function-driven Shape Matching

Fine localization for the road markings within the candidate bounding box was formulated as the determination of the 6D vector of the specific road marking template, including the location (x, y, z), orientation (ϕ, θ), and scale s, where ϕ is the angle between the road marking plane and the horizontal plane in the raw point cloud coordinate system, and θ is the angle between the principal direction of the template and the height direction of the bounding box. Twelve kinds of road marking templates of interest were constructed according to the Chinese national standards for road traffic signs and markings (General Administration of Quality Supervision and China 2009).

The shape matching energy function consists of two parts, as can be seen in Eq. (11.18): an overall description of the intensity distribution within a road marking, and local distribution of intensity gradients along the boundary of the road marking template. By maximizing the energy function, the optimal location, orientation, and scale are acquired while the road marking is segmented and modeled precisely.

$$E_{\text{fine}}\left(x,y,z,\phi,\theta,s\right)=\alpha*\frac{1}{N}\sum_{i=1}^{N}I_i+\left(1-\alpha\right)*\frac{1}{N_e}\sum_{j=1}^{N_e}G_j \tag{11.18}$$

where α is the weight, I_i is the intensity of the i-th point in the template, N is the count of points located in the template, N_e is the number of the sample points on the border of the template, and G_j is the intensity gradient of the j-th border point, which is computed by Eq. (11.19).

$$G=\left|\frac{1}{N_{\text{in}}}\sum_{i=1}^{N_{\text{in}}}I_i-\frac{1}{N_{ex}}\sum_{i=1}^{N_{\text{ex}}}I_i\right| \tag{11.19}$$

where N_{in} is the interior point count and N_{ex} is the exterior point count within a searching radius. The searching radius is twice the average nearest point distance within the bounding box.

Road markings at the sectioning line are incomplete and are sometimes predicted in the incorrect category by the object detection network. Moreover, several road markings are geometrically similar. For instance, the left turn and right turn arrows are mirror-symmetrical in geometry, causing errors in category prediction by the detection network. Hence, the bounding box prediction confidence E_{coarse} and the localization score E_{fine} are combined to rerank the candidate intersecting road markings in descending order.

$$\text{E}=\sigma E_{\text{coarse}}+\left(1-\sigma\right)E_{\text{fine}}^{n} \tag{11.20}$$

Specifically, the candidate reranking with optimal selection occurs when (1) overlap between the candidate road markings exists, or (2) the candidate road marking has a mirror-symmetric twin. Figure 11.7 illustrates the procedure of candidate reranking with optimal selection.

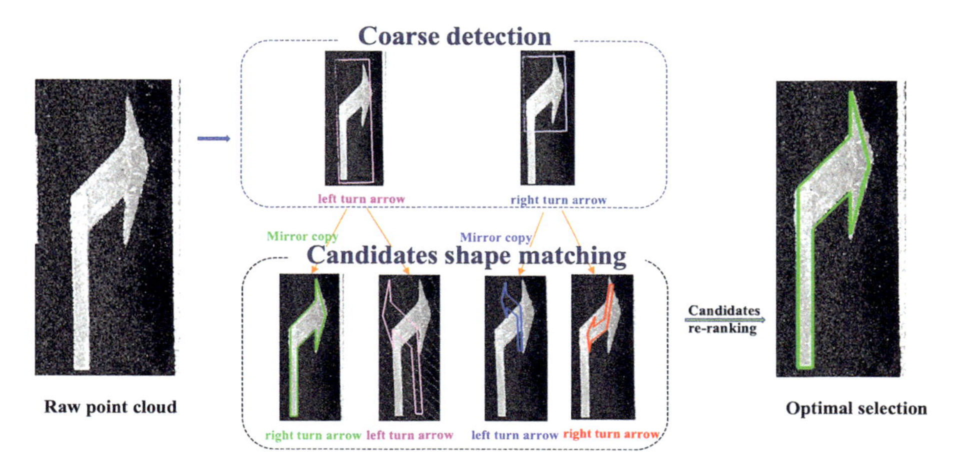

FIGURE 11.7 Illustration of fine localization and modeling (Mi et al. 2021).

11.2 EXPERIMENTS

11.2.1 Dataset Description

Three datasets collected by an Alpha3D mobile system are used to verify the effectiveness of the proposed method. The Alpha3D system is equipped with a long-range and precise RIEGL VUX-1 laser scanner, a high precision inertial measurement unit (IMU), and an advanced global navigation satellite system (GNSS) receiver in combination with a high-resolution high dynamic range (HDR) panoramic camera. The relative accuracy of the acquired point cloud is about 8mm within the range of 100m from the scanner to the point. Dataset I covers an industrial area, an urban area, and an expressway. Dataset II is collected in a complex urban area with many worn and incomplete road markings. Table 11.1 lists the detailed description of two datasets while Figure 11.8 shows the corresponding vehicle's trajectories.

Road markings of interest in the three datasets were manually annotated. There were 13 semantic types, including seven types of arrows (straight arrow, left turn arrow, right turn arrow, straight & left turn arrow, straight & right turn arrow, left & U-turn arrow, and U-turn arrow), forbidden, diamond, stop line, crosswalk, dashed lane, and background. Dataset I was selected as the training dataset, and datasets II and III served as test datasets.

TABLE 11.1
Quantitative Dataset Description

Dataset	Data Size	Road Length	Coverage Area
I	79.53 GB	36.467 km	1.4 km*2.0 km
II	53.80 GB	28 km	1.4 km*1.5 km

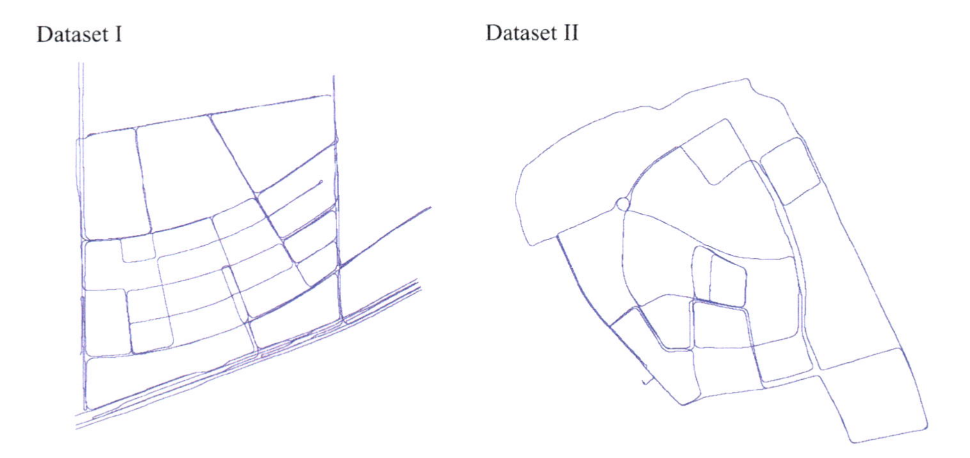

FIGURE 11.8 Vehicle's trajectories of two datasets.

TABLE 11.2
Numbers of Labeled Road Markings for Training

Category	Straight Arrow	Left Turn Arrow	Right Turn Arrow	Straight & Left Arrow	Straight & Right Arrow	U-turn Arrow
Quantity	1724	1485	679	510	904	749
Category	**Forbidden**	**Diamond**	**Left & U-turn Arrow**	**Stop Line**	**Crosswalk**	**Dashed Lane**
Quantity	731	149	160	1048	14340	19739

Feature maps were generated from the labeled point clouds after sectioning according to the proposed method. Because the number of samples in the training set was limited, data augmentation was carried out through coordinates translations, rotations around the z-axis, and partial occlusion. After data augmentation, 4348 feature maps were available in the training dataset. Table 11.2 lists the number of road markings of each type in the training dataset after data augmentation.

11.2.2 Experimental Setting

The parameters for road boundary extraction and vectorization were listed in Table 11.3.

Regarding road marking extraction, feature map resolutions in the height and width direction were 0.08 m and 0.03 m, respectively. The hyper-parameters in YOLOv3 were fine-tuned through repeated experiments. According to the approximate feature map size in the dataset, the input image size was set to 640 by 800 in terms of width and height, respectively. The batch size was set to 64, and the learning

TABLE 11.3
Parameter Settings for Road Boundary Extraction and Vectorization

Parameters	Description	Values
Δh	Height difference threshold in Ground filtering	0.2 m
G_r	Grid resolution	0.2 m
w_{xy}	Weight on horizontal distance in supervoxel-generation	0.25
w_z	Weight on height difference in supervoxe-generation	0.15
w_d	Weight on point density in supervoxel-generation	0.6
r	Horizontal searching radius of point density for sole point	0.03 m
s	Forward step size in the Kalman filter	0.1 m
α	Weighting parameter in false positive removal	4

TABLE 11.4
Parameter Settings for Road Marking Extraction and Modeling

Parameters	Description	Values
τ	NMS threshold in global intraclass NMS	0.5
ε	Weight of bounding box width for lines/others in global intraclass NMS	0.1/0.5
α	Weight of average intensity inside the template in the energy function	0.5
σ	Weight of coarse localization confidence in candidates reranking	0.2

rate was 0.001. Other parameters were set to their default values. Table 11.4 gives the descriptions and settings of vital parameters.

11.2.3 Experimental Results

11.2.3.1 Road Boundary

Vectorized road boundaries of the datasets I and II using the proposed method are shown in the bottom right of Figures 11.9 and 11.10 respectively. And the raw MLS point clouds of the two datasets are shown in the top right of the Figures 11.9 and 11.10, respectively. Besides, several typical parts are also enlarged for visual inspection. The results of road boundaries on the straight and curved roads are shown in Figures 11.9(a), (b), and 11.10(a), (b), respectively. Besides, road boundaries of the complex intersections are well extracted and vectorized as illustrated in Figures 11.9(c), (d), and 11.10(c), (d). Even for some challenging scenes, such as curbs in various irregular shapes in Figure 11.9(e), boundaries surrounded with the fence in Figure 11.9(f), and wall-like road boundaries in Figure 11.10(e), road boundaries are extracted and vectorized accurately and smoothly. It is also observed that road boundaries are not only been extracted and vectorized accurately and smoothly, but the most missing parts caused by the moving objects occlusion and non-motor access are completed after tracking, as shown in Figure 11.10. Figure 11.10(f)

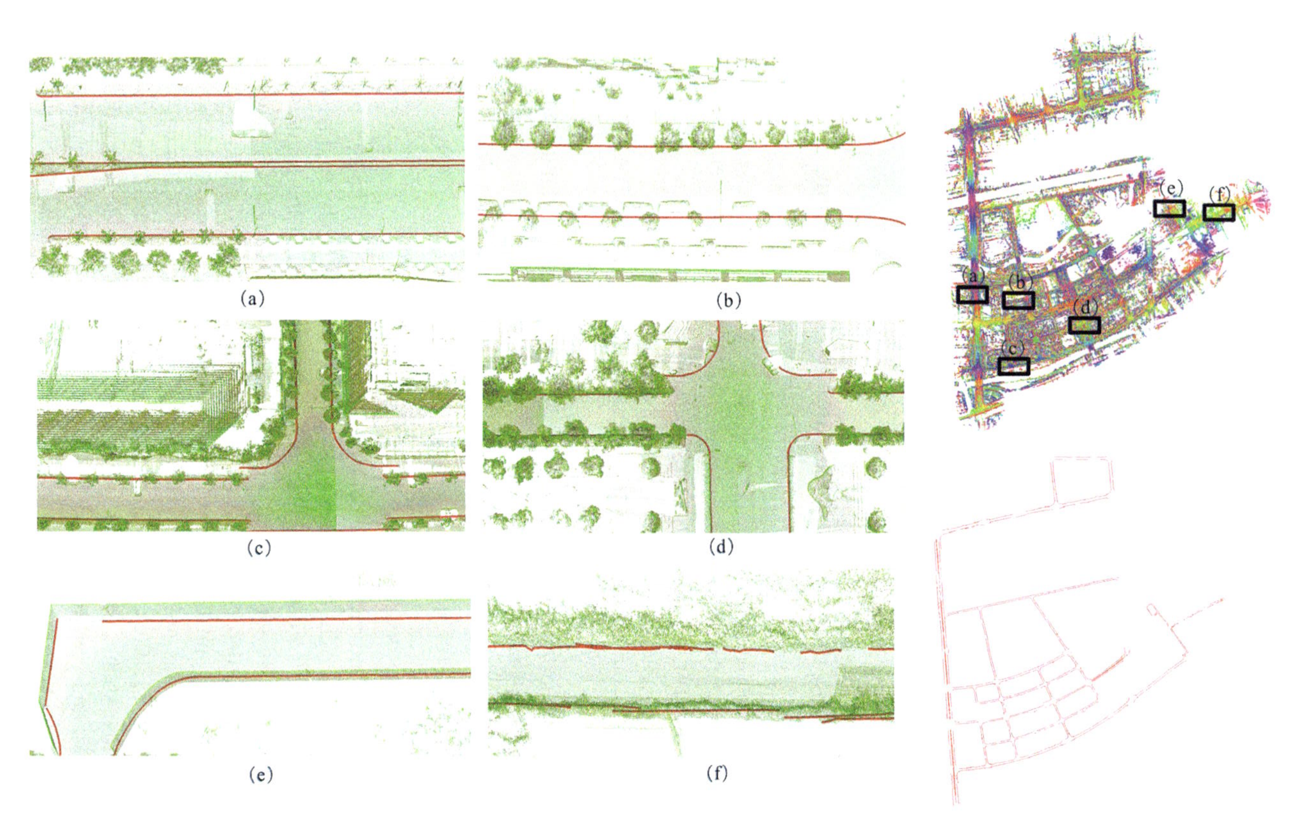

FIGURE 11.9 Vectorized road boundary on Dataset I by the proposed method (Mi et al. 2022).

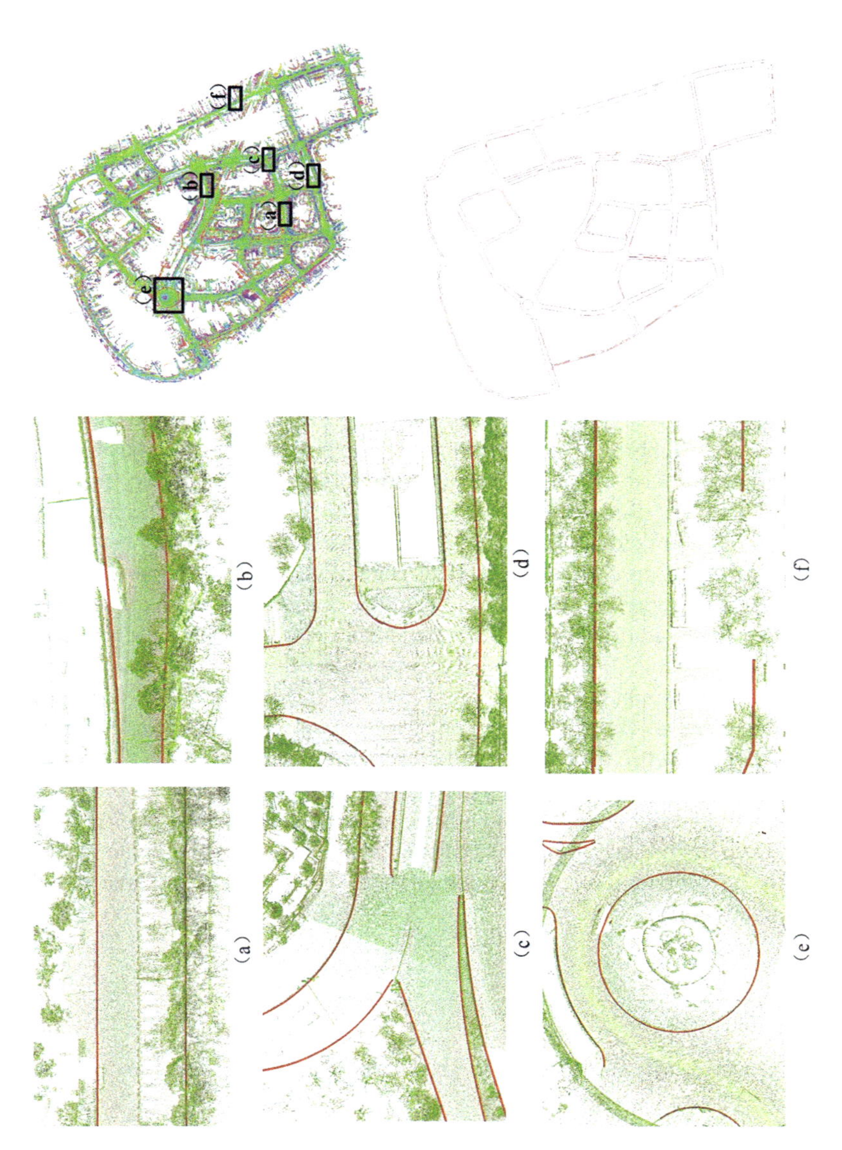

FIGURE 11.10 Vectorized road boundary on Dataset II by the proposed method (Mi et al. 2022).

shows the road boundary completion for the missing parts caused by the vehicle. However, in cases where there is heavy occlusion along roads or no structured road boundary without a specific elevation step, our method may fail to extract the correct road boundary. For instance, road boundaries are not completed well as too much occlusion by cars along roadsides for cases in Figure 11.9(f). As for cases in Figure 11.10(f), our method failed to extract the continuous and smooth road boundary since the road here is without a specifically structured boundary but natural vegetation.

To quantitatively evaluate the performance of the proposed method, we selected more than 6.0 km of road boundary data from each dataset. The ground truth road boundary was extracted from original point clouds, and missing parts caused by occlusion of moving objects on the road and non-motor access along the road were manually completed. And it is stored in the form of polylines, which is discretized into ordered point sets with a distance of 5 cm for quantitative evaluation. Similarly, the final vectorized road boundary is also discretized into ordered point sets with a distance of 5 cm for quantitative evaluation. When the point of extracted road boundary exists around the ground truth within the 5 cm buffer, it is considered as TP (True Positive) point, otherwise, it is FP (False Positive) point. The quantitative evaluation results are listed in Table 11.5, which shows that the proposed method has high computing efficiency in terms of producing 1km vectorized road boundary per minute according to time cost. Because only the points in the uneven grids of a small proportion in raw MLS point clouds are taken into consideration and normal estimation is never adopted in the process of supervoxel-generation. And the superiority of the accurate vectorized representation of road boundary rather than point clouds are: (1) the data size of the vectorized road boundary achieves a significant reduction compared with dense points. For instance, the data size of the extracted road boundary points from original point clouds is about 47.70 MB but only 0.28 MB after vectorization, which achieves about 170 times data compression rate for the selected evaluation data from dataset II. Similarly, the data compression rate is up to 319 times for the selected data from dataset I; (2) the road boundary represented as the multiple continuous cubic Bezier curves is smoother compared with the polylines connected by extracted points directly, which is critical for mapping; (3) the parametric road boundary is utilized to track road boundaries to boost the completeness of the extracted road boundary.

TABLE 11.5
Quantitative Evaluation of the Vectorized Road Boundaries on Two Datasets

						Data Size (MB)	
Dataset	GT(m)	TP(m)	FP(m)	FN(m)	Time Cost(s)	Points	Vector
I	6372.14	6255.96	231.89	116.18	444.23	35.10	0.11
II	6987.83	6586.82	216.58	401.01	486.13	47.70	0.28

11.2.4 Road Marking

Figure 11.11 shows qualitative illustrations of the modeled road markings using the proposed approach from datasets II, where the global overviews are displayed on the top left and several typical road stretches are enlarged for detailed inspection. Figure 11.12 shows road marking extraction and modeling results for three complex road stretches with different road markings scales, incomplete raw data, and various intensities and point densities.

The quantitative evaluation metrics for road marking extraction are precision, recall, $F1$-score, and mIoU. TP, FP, FN are the number of true positives in the extracted results that IoU between the extracted road markings and corresponding ground truth are higher than 0.5, the number of false positives in the extracted results, and the number of false negatives in the ground truth, respectively.

$$m\mathrm{IoU}_C = \frac{\sum_{i=1}^{N} A_i^{M \cap G}}{\sum_{i=1}^{N} A_i^{M} + \sum_{i=1}^{N} A_i^{G} - \sum_{i=1}^{N} A_i^{M \cap G}}$$

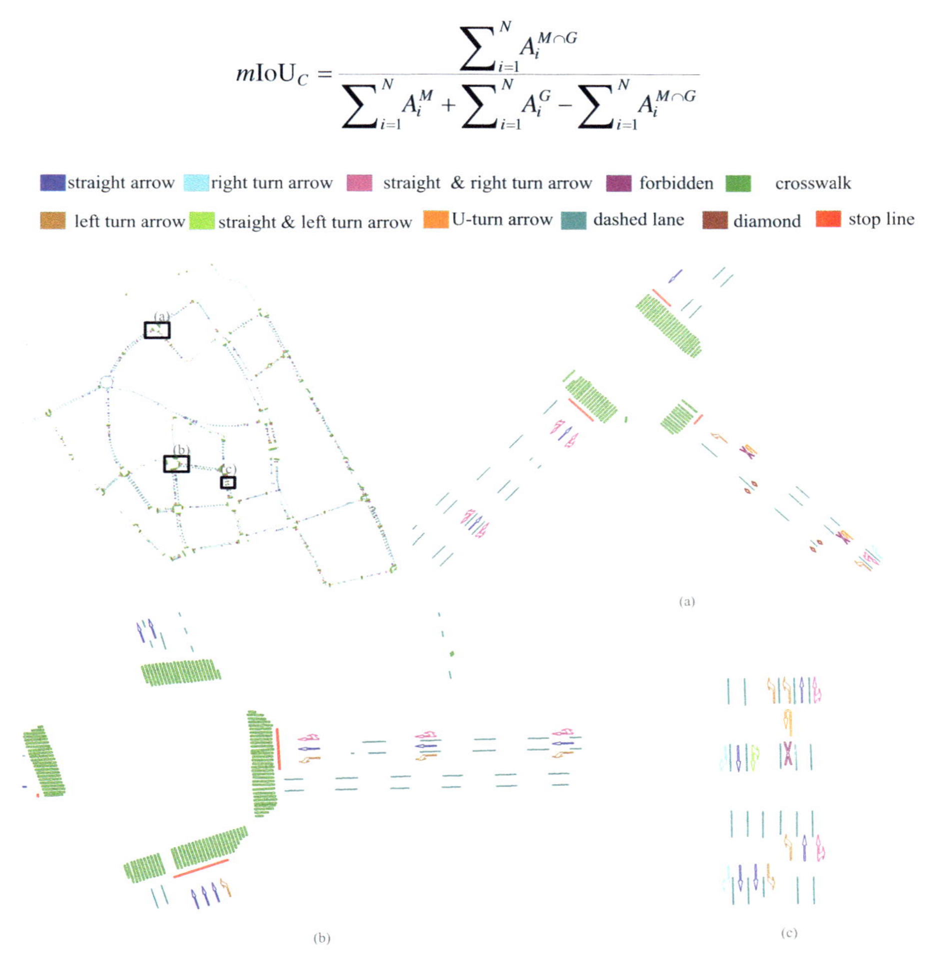

FIGURE 11.11 Road marking modeling results on Dataset II by the proposed method (Mi et al. 2021).

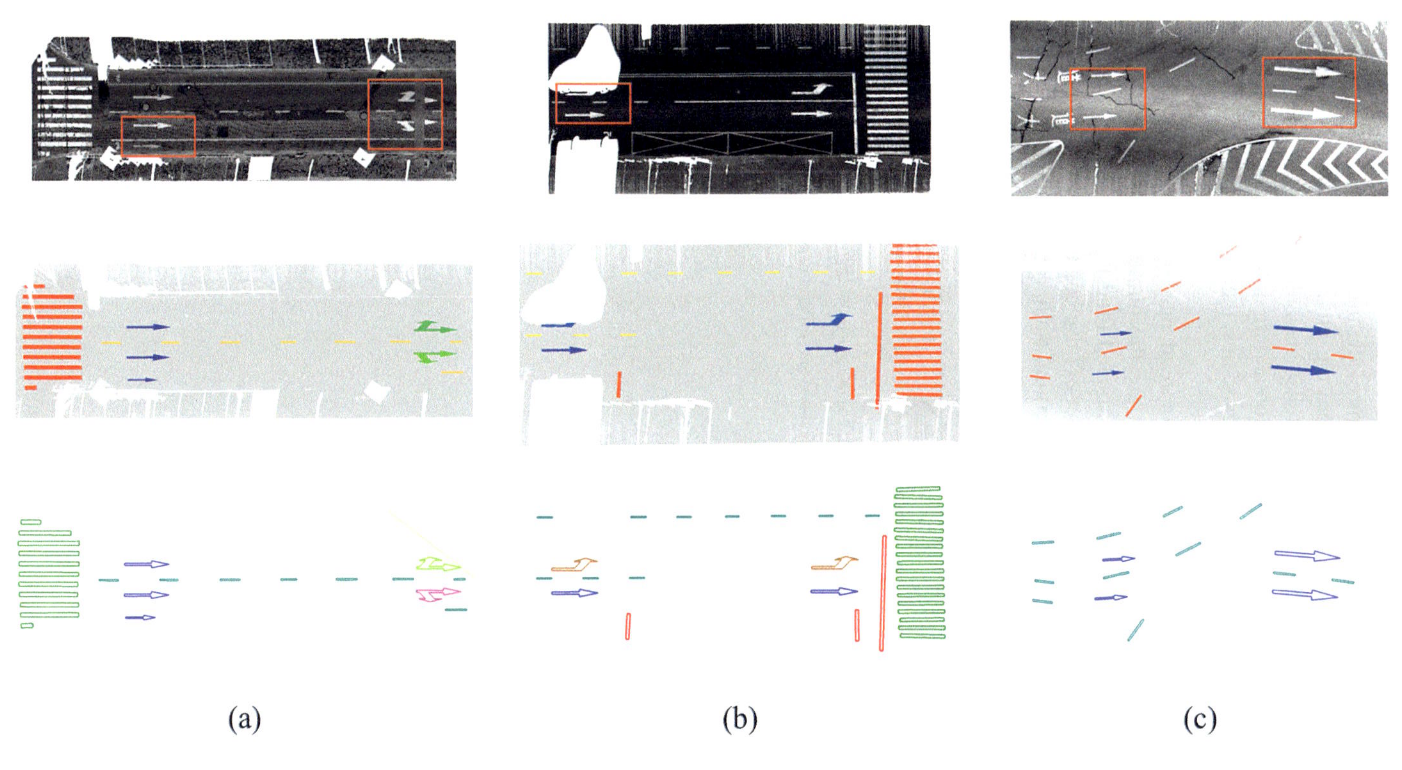

FIGURE 11.12 Road marking extraction and modeling results on challenging scenes, where road markings are with different scales and incomplete (Mi et al. 2021).

TABLE 11.6
Quantitative Evaluation for Road Marking Extraction Results on Dataset II

Category	GT	TP	FP	FN	Recall	Precision	$F1$-score
Straight arrow	218	211	23	7	0.968	0.902	0.934
Left turn arrow	107	99	4	8	0.925	0.961	0.943
Right turn arrow	105	95	6	10	0.905	0.941	0.922
Straight & left arrow	28	27	4	1	0.964	0.871	0.916
Straight & right arrow	71	63	2	8	0.887	0.969	0.927
U-turn arrow	17	15	0	2	0.882	1	0.937
Forbidden	11	10	4	1	0.909	0.714	0.800
Diamond	70	70	0	0	1.000	1.000	1.000
Stop line	108	100	6	8	0.926	0.943	0.935
Crosswalk	2240	2035	88	205	0.908	0.959	0.933
Dashed lane	2262	2107	112	155	0.931	0.950	0.940
Overall	5237	4832	249	405	0.923	0.951	0.937

TABLE 11.7
Quantitative Evaluation for Road Marking Modeling Results

Category	Straight Arrow	Left Turn Arrow	Right Turn Arrow	Straight & Left Arrow	Straight & Right Arrow	U-turn Arrow
$mIoU_c$	0.755	0.809	0.848	0.804	0.825	0.567
Category	**Forbidden**	**Diamond**	**Left & U-turn arrow**	**Crosswalk**	**Stop line**	**Dashed lane**
$mIoU_c$	0.504	0.538	0.690	0.721	0.712	0.784

where A_i^M is the area of the i-th true positive road marking model in class C, A_i^G is the area of the corresponding ground truth model, and N is the count of the true positive road marking models in class C.

The quantitative evaluations for road marking extraction results on datasets II are listed in Table 11.6. The quantitative evaluations for road marking modeling results are listed in Table 11.7.

11.3 ANALYSIS AND DISCUSSION

11.3.1 Road Boundary

Table 11.5 shows that the proposed method has high computing efficiency in terms of producing 1 km vectorized road boundary per minute according to time cost. Because only the points in the uneven grids of a small proportion in raw MLS point clouds are taken into consideration and normal estimation is never adopted in the process of supervoxel-generation. And the superiority of the accurate vectorized representation of road boundary rather than point clouds are: (i) the data size of the vectorized road boundary achieves a significant reduction compared with dense points. For instance, the data size of the extracted road boundary points from original point clouds is about 47.70 MB but only 0.28 MB after vectorization, which achieves about 170 times data compression rate for the selected evaluation data from dataset II. Similarly, the data compression rate is up to 319 times for the selected data from dataset I; (ii) the road boundary represented as the multiple continuous cubic Bezier curves are smoother compared with the polylines connected by extracted points directly, which is critical for mapping; (iii) the parametric road boundary is utilized to track road boundaries to boost the completeness of the extracted road boundary.

11.3.2 Road Marking

Road marking extraction and modeling results demonstrate that the proposed top-down method is robust to the variations in intensity and point density, and incomplete raw data. Existing bottom-up methods for the road marking extraction from point clouds take local intensity information to extract the foreground points or pixels, a process that depends heavily on intensity uniformity. However, wear on road markings and the scanner mechanism can lead to large variations in road marking intensity, which in turn leads to challenges in extracting a complete set of road marking points. In addition, occlusion caused by the vehicles on the road causes incomplete raw data. These incomplete road marking segments are challenging for bottom-up methods to classify and model correctly and precisely. Instead of the bottom-up strategy, the proposed top-down approach used a general object detection network to detect the candidate bounding boxes with corresponding semantic labels. An energy function-driven shape matching procedure was implemented to locate the road markings with precise positions, orientations, and scales. By taking both detection confidence and shape matching score into account, the candidate road markings were reranked with optimal selection to produce the road marking models. The proposed top-down method considers the geometric shapes and radiometric appearance of road markings in the whole pipeline, a process that is less sensitive to the imperfect raw data than bottom-up methods.

As listed in Table 11.6, the proposed approach achieved a performance of 92.3% and 95.1% in terms of overall recall and precision for road marking extraction. As listed in Table 11.7, the *mIoU* achieved a performance of 0.504 for road marking modeling. Among the 12 kinds of road markings, the modeling precisions for the forbidden, diamond, and U-turn arrow were lower than those of the others, which was mainly caused by shape differences between the tested data and the national standard templates.

11.4 CONCLUSION

To address the challenges in precise extraction and structural modeling of road elements, we propose the following key topics: (1) MLS point clouds are characterized by disorder, discreteness, and large volumes of data. However, the size of road boundaries in massive outdoor scene point clouds is relatively tiny. Aiming at the structural modeling of road boundaries, we propose a road boundary extraction and vectorization approach from MLS point clouds, combining supervoxel-generation and adaptively structural modeling. The supervoxel-generation procedure converts the sparse and unordered points into the supervoxels with fine boundaries, which is the basis of the following curbs' recognition and modeling. A trajectory-guided distance contraction clusters the adjacent supervoxels into initial road boundary entities. An adaptive parameterized representation and a tracking algorithm are proposed to produce smooth, complete, and continuous vectorized road boundaries. The vectorized road boundaries' precision and recall on the experimental dataset are better than 96.4% and 94.3%, respectively. (2) Facing the challenges of imperfect point cloud quality and the various sizes and shapes of road markings on road surfaces, a two-stage, coarse-to-fine road marking detection and reconstruction framework is proposed in this chapter. This framework consists of coarse detection and fine localization stages. Coarse detection mainly focuses on detecting candidate bounding boxes with road marking semantics. While fine localization stage formulates the road marking reconstruction into the energy function-guided template matching problem. Road marking segmentation and reconstruction is achieved simultaneously by determining the template's localization, orientation, and scale parameters. Furthermore, the optimal road marking models are selected by reranking the combined confidences that consider both coarse detection and fine localization scores. The extraction precision and recall on the urban dataset are better than 95.1% and 92.3%, respectively. The reconstructed road marking models' mean IoU is better than 0.504.

REFERENCES

Forrest, A.R. 1980. "The Twisted Cubic Curve: A Computer-Aided Geometric Design Approach." *Computer-Aided Design* 12 (4): 165–72.

General Administration of Quality Supervision, Inspection, and Quarantine of the People's Republic of China. 2009. "GB5768.3-2009: Road Traffic Signs and Markings. Part 3: Road Traffic Markings."

Hershberger, John Edward, and Jack Snoeyink. 1992. "Speeding up the Douglas-Peucker Line-Simplification Algorithm." University of British Columbia, Department of Computer Science.

Holgado-Barco, A., D. Gonzalez-Aguilera, P. Arias-Sanchez, and J. Martinez-Sanchez. 2015. "Semiautomatic Extraction of Road Horizontal Alignment from a Mobile LiDAR System." *Computer-Aided Civil and Infrastructure Engineering*. doi:10.1111/mice.12087

Li, Y., W. X. Wang, S. J. Tang, D. L. Li, Y. K. Wang, Z. L. Yuan, R. Z. Guo, X. M. Li, and W. Q. Xiu. 2019. "Localization and Extraction of Road Poles in Urban Areas from Mobile Laser Scanning Data." *Remote Sensing*. 401. doi:10.3390/rs11040401

Lin, Tsung-Yi, Piotr Dollár, Ross Girshick, Kaiming He, Bharath Hariharan, and Serge Belongie. 2017. "Feature Pyramid Networks for Object Detection." *Proceedings of the IEEE Conference on Computer Vision and Pattern Recognition*, 2117–25.

Liu, Li, Wanli Ouyang, Xiaogang Wang, Paul Fieguth, Jie Chen, Xinwang Liu, and Matti Pietikäinen. 2020. "Deep Learning for Generic Object Detection: A Survey." *International Journal of Computer Vision* 128 (2): 261–318.

Mi, Xiaoxin, Bisheng Yang, Zhen Dong, Chi Chen, and Jianxiang Gu. 2022. "Automated 3D Road Boundary Extraction and Vectorization Using MLS Point Clouds." *IEEE Transactions on Intelligent Transportation Systems* 23 (6): 5287–97. doi:10.1109/TITS.2021.3052882

Mi, Xiaoxin, Bisheng Yang, Zhen Dong, Chong Liu, Zeliang Zong, and Zhenchao Yuan. 2021. "A Two-Stage Approach for Road Marking Extraction and Modeling Using MLS Point Clouds." *ISPRS Journal of Photogrammetry and Remote Sensing* 180: 255–68. doi:10.1016/j.isprsjprs.2021.07.012

Rastiveis, Heidar, Alireza Shams, Wayne A Sarasua, and Jonathan Li. 2020. "Automated Extraction of Lane Markings from Mobile LiDAR Point Clouds Based on Fuzzy Inference." *ISPRS Journal of Photogrammetry and Remote Sensing* 160: 149–66.

Redmon, Joseph, and Ali Farhadi. 2018. "Yolov3: An Incremeental Improvement." *arXiv Preprint arXiv:1804.02767.*

Welch, Greg, and Gary Bishop. 1995. "An Introduction to the Kalman Filter."

Part V

Software and Applications

12 Point Cloud Processing Software – Point2Model

12.1 ARCHITECTURE OF POINT2MODEL

As shown in Figure 12.1, the overall architecture of the software is divided into four parts: database, data access layer, business logic layer, and presentation layer.

The database is the foundation layer of the whole architecture and is responsible for storing heterogeneous data from multiple sources such as point cloud, image, and vector.

As a bridge between the Database and the Business Logic Layer, the Data Access Layer functions as a data retrieval, data distribution, data reading and writing, log output, and exception combing layer. It is responsible for interacting with the Database and providing a unified data access interface for the Business Logic Layer. The design of the Data Access Layer is crucial to ensuring the efficient transmission and processing of data.

The Business Logic Layer is responsible for handling the core business logic of the software, which is the "brain" of the software, including modules for task distribution, data preprocessing, quality control, data fusion, target extraction, quality checking, etc. The Business Logic Layer obtains the required data from the Data Access Layer, performs the necessary processing and calculations, and then passes the results to the Presentation Layer. The flexibility and extensibility of the Business Logic Layer are crucial for adapting to the ever-changing business needs.

The Presentation Layer interacts directly with the user, presenting the results of the Business Logic Layer to the user intuitively through functions such as data visualization and data roaming, as well as providing an interface for the user to enter data and execute commands. Data visualization makes complex data and analysis results easy to understand, while the user interface ensures the ease of use and accessibility of the software. The data interaction component ensures that users can effectively communicate with the system in both directions, input their needs, and get feedback from the system. The design of the presentation layer directly affects the user experience and ease of use of the software.

Throughout the architecture, the layers are closely linked and organized. The Data Layer provides the upper layers with raw data and essential data manipulation functions; the Data Access Layer further processes the data and converts it into information that the Business Logic Layer can use; the Business Layer performs complex business processing and generates valuable business results; and the Presentation Layer displays these results to the users in graphical form, and at the same time collects feedback and input from the users. This multi-tier architecture allows the Point2Model software to maintain high efficiency and stability when

DOI: 10.1201/9781003486060-17

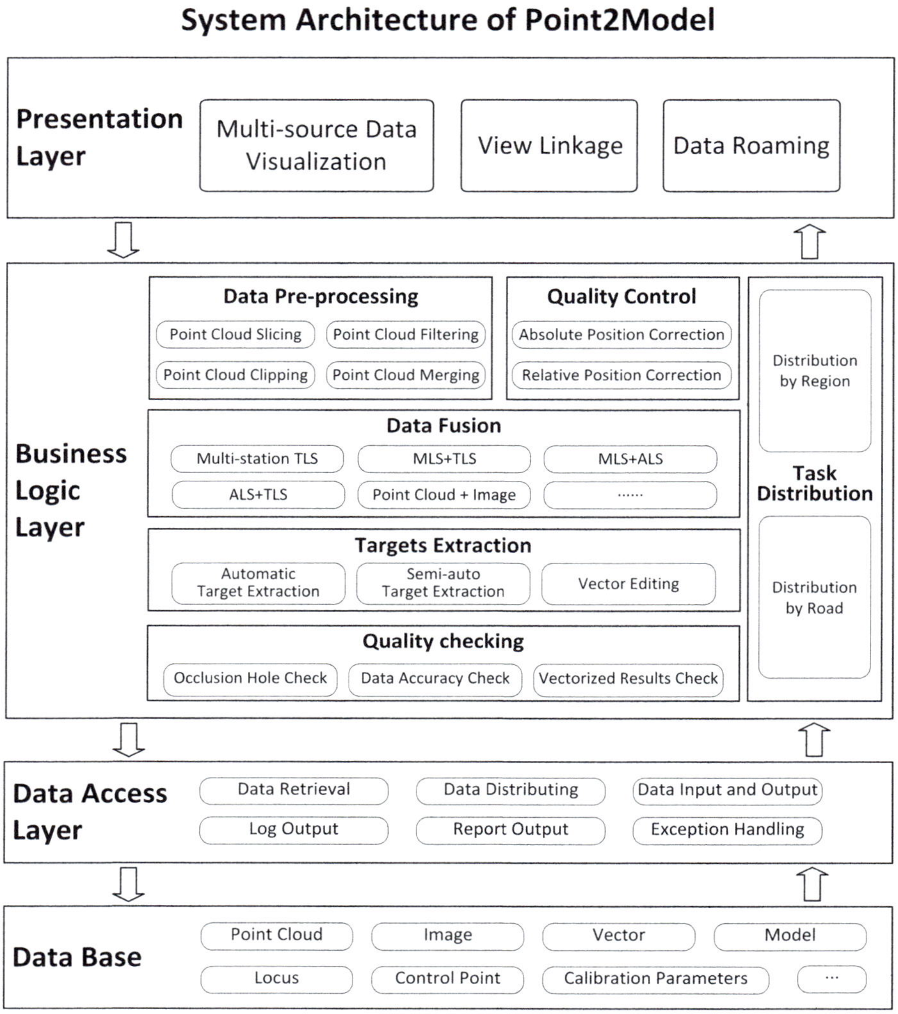

System Architecture of Point2Model

Presentation Layer
Multi-source Data Visualization
View Linkage
Data Roaming

Business Logic Layer
Data Pre-processing
Point Cloud Slicing
Point Cloud Filtering
Point Cloud Clipping
Point Cloud Merging
Quality Control
Absolute Position Correction
Relative Position Correction
Data Fusion
Multi-station TLS
MLS+TLS
MLS+ALS
ALS+TLS
Point Cloud + Image
......
Targets Extraction
Automatic Target Extraction
Semi-auto Target Extraction
Vector Editing
Quality checking
Occlusion Hole Check
Data Accuracy Check
Vectorized Results Check
Task Distribution
Distribution by Region
Distribution by Road

Data Access Layer
Data Retrieval
Data Distributing
Data Input and Output
Log Output
Report Output
Exception Handling

Data Base
Point Cloud
Image
Vector
Model
Locus
Control Point
Calibration Parameters
...

FIGURE 12.1 The system architecture of Point2Model.

processing large-scale point cloud data and also facilitates subsequent maintenance and expansion.

12.2 COMPONENTS OF POINT2MODEL

Point2Model includes five functional modules, including (1) Data Organization Management and Visualization Module, (2) Point Cloud Position Accuracy Improvement Module, (3) Multi-source Data Fusion Module, (4) Target Extraction Module, and (5) Model Reconstruction Module.

12.2.1 Data Organization Management and Visualization Module

Data Organization Management and Visualization Module is based on a self-developed high-efficiency index structure of massive point cloud data, with index construction efficiency better than 40 seconds/GB, the response time of TB-level point cloud display less than two seconds, and support for fast roaming of accessible viewpoints; it can efficiently manage various types of heterogeneous data, such as point clouds, images, vectors, trajectories, control points, calibration parameters, and target extraction results, etc., through the structure of tree-like layers.

12.2.2 Point Cloud Position Accuracy Improvement Module

The Point Cloud Position Accuracy Improvement Module can utilize the point cloud repeated observation information and historical high-precision control information to realize large-scale vehicle, airborne, and backpack point cloud position accuracy improvement, and the accuracy is better than 5 cm after correction. The relative position improvement of the vehicle point cloud without control constraints is achieved by designing a vehicle point cloud position improvement method based on the vehicle point cloud trajectory segmentation combined with the two-by-two global consistency of the coarse-to-fine vehicle point cloud alignment. The process is based on the segmentation of the point cloud trajectory. Firstly, adaptive segmentation of the on-board point cloud trajectory considers the distribution characteristics of the point cloud error. Then, the point cloud is divided into small segments of different lengths. This project proposes a point cloud matching algorithm based on the BSC feature descriptor, which combines the small segments into longer coarse segments, provides a good initial value for the two-by-two alignment of coarse segments to provide a good initial value for the two-by-two alignment of fine segments, and performs the unreliable matching relationship after obtaining the results of two-by-two alignment; and finally, by constructing a graph model of the bit positions of the point clouds of different segments, and redistributing the on-board point cloud segmentation errors utilizing graph optimization, which distributes errors evenly to all redistributed point clouds. The error is evenly distributed to all the revisited segments, and finally, the optimized average relative accuracy is made to reach the centimeter level. As shown in Figure 12.2, after the point cloud relative position accuracy improvement, the point cloud position quality is significantly improved.

12.2.3 Multi-source Data Fusion Module

The Multi-source Data Fusion Module realizes fully automatic and uniform positioning, attitude fixing, and fusion of diversified scenes (e.g., city, forest, mountain, tunnel, ocean, etc.) and heterogeneous data from multiple sources (e.g., satellite-borne, airborne, vehicle-mounted, ground station, vehicle-mounted image, etc.) by overcoming the identification and pairing of consistent features of the same feature and the high-precision solving model of the unified spatial and temporal bases. Specifically, Point2Model realizes the alignment and fusion between multi-platform point clouds, such as vehicle-mounted point clouds and panoramic images, multi-view ground

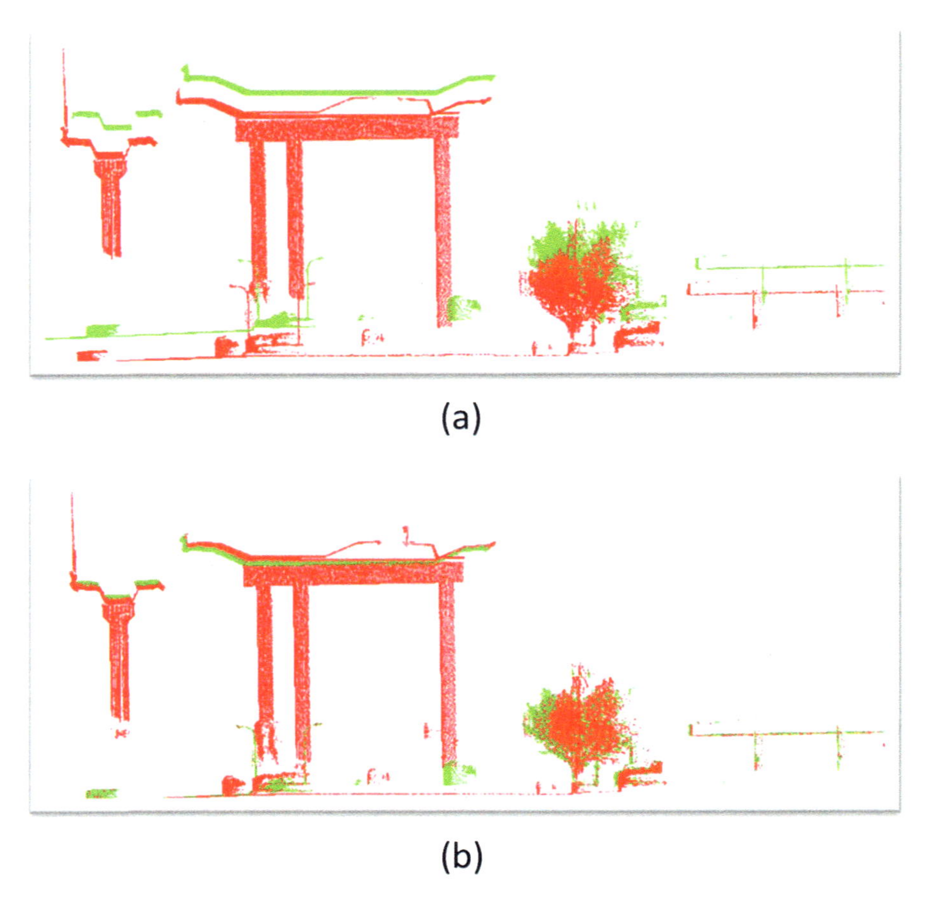

FIGURE 12.2 (a) Before the improvement of the relative positional accuracy of the point cloud; (b) After the improvement of the relative positional accuracy of the point cloud.

station point clouds and vehicle-mounted point clouds, multi-view ground station point clouds and airborne point clouds, airborne point clouds and vehicle-mounted point clouds, and hand-held scanner point clouds. Figure 12.3 shows the stacked point cloud and image after data fusion.

12.2.4 Target Extraction Module

The Target Extraction Module utilizes the urban feature element segmentation and 3D extraction method with integrated feature engineering and deep learning to realize automatic extraction and vectorization of more than 30 types of feature elements in urban and high-speed scenes, complete semantic understanding and structured expression of the extracted targets, and realize the leap from visual expression to computable analysis. Specifically, Point2Model realizes the automated 3D extraction

FIGURE 12.3 Stacked point cloud and image after registration.

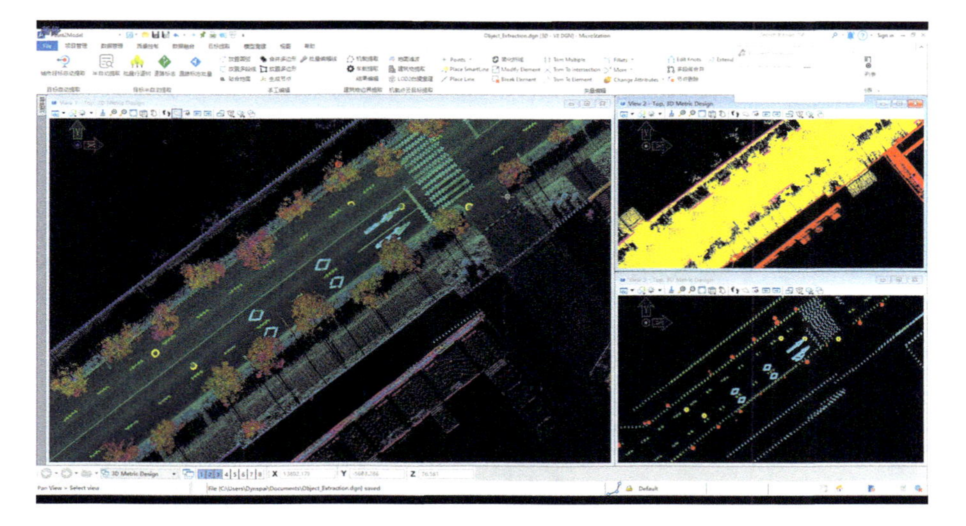

FIGURE 12.4 Fully automatic target extraction results.

of typical road facilities such as street trees, road boundary lines, road traffic signs, buildings, utility poles, street lights, traffic signs, guardrails, and so on, based on machine learning and deep learning. Figure 12.4 displays some of the fully automatic target extraction results.

However, the extraction results of automated algorithms do not guarantee full correctness and still need to be checked and corrected manually. Therefore, as shown in Figure 12.5, Point2Model has designed a convenient semi-automatic interactive target extraction function for users, who only need to use the mouse to click or box the target to get the accurate 3D spatial location and contour of the target of interest.

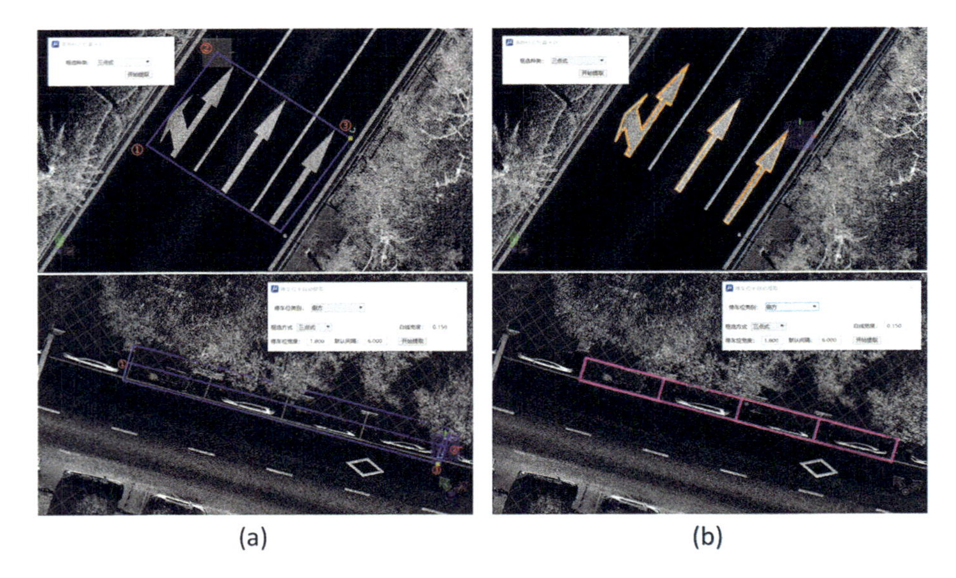

FIGURE 12.5 (a) Range of interest boxed by the user; (b) Semi-automatic target extraction results (traffic signs and parking spots).

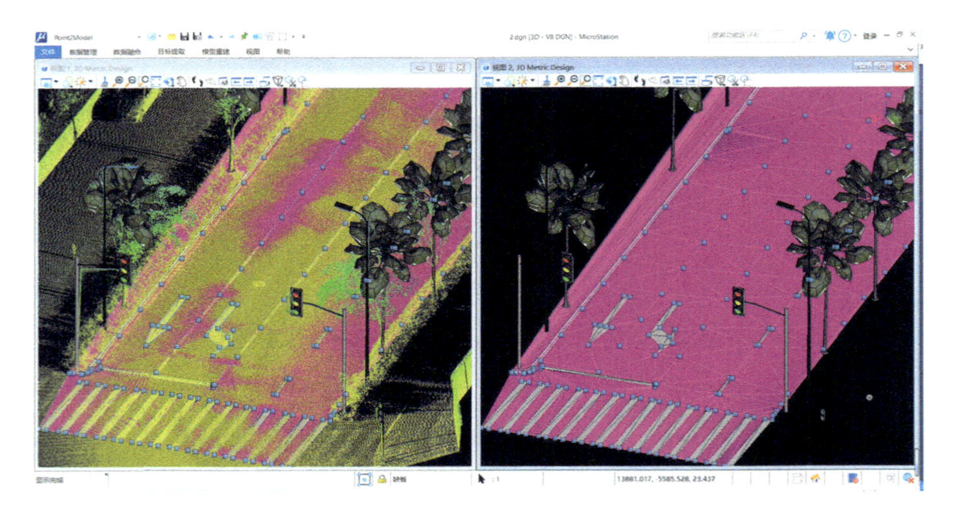

FIGURE 12.6 Road model reconstructed by the Model Reconstruction Module.

12.2.5 Model Reconstruction Module

The Model Reconstruction Module completes the parametric expression of roads, forest trees, traffic facilities, buildings, and other elements through the core technologies of geometric primitive extraction, attribute map construction, and model enhancement and realizes the reconstruction of high-precision and structured 3D models. Figure 12.6 presents a road model reconstructed by the Model Reconstruction Module.

13 Applications of Ubiquitous Point Cloud

13.1 INTELLIGENT TRANSPORTATION SYSTEMS

Intelligent Transportation System (ITS) is an important part of modern urban traffic management and control, which integrates advanced information technology, communication technology, sensor technology, electronic control technology, and computer technology into the entire transportation management system, aiming to improve traffic efficiency, reduce congestion, enhance safety, and decrease environmental impact. With the rapid development of intelligent transportation systems and automatic driving technology, its necessary component - HD Map (High-Definition Map), is considered a key part of future transportation, which is a significant supplement to real-time perception of transportation scenarios in all space and time and the basis of the whole process of operation and control of transportation. All functions above require high geometric accuracy and high freshness. Compared with traditional maps, HD Map provides road network information, containing extremely detailed data on road alignments, traffic signs, and lane information, such as lane widths, road curvature, and gradient. Such information is critical for supporting assisted-/self-driving vehicles in intelligent transportation systems, improving traffic flow management, optimizing route planning, and enhancing road safety.

Ground survey data have played a crucial role in the history of map production over four centuries. The earliest ground surveys relied on simple hand measurements and visual estimations, but over time, map production techniques have evolved, including using more accurate measurement tools and methods. Photogrammetric surveying methods emerging in the 1930s eased the laborious task of field surveying. In the 1980s, automated mapping systems consisting of total stations and digital mapping further liberated productivity and greatly improved the accuracy and reliability of maps. With the development of aerial photography in the early 20th century, aerial pictures provided valuable perspectives and information for maps, and remote sensing technology began to be applied to map production. With the development of satellite technology, the acquisition of remote sensing data gradually shifted to satellite imaging. In the 1960s, with the launch of the first Earth observation satellite, remote sensing technology entered a new era. From the 1970s to the 1990s, with the launch of more advanced satellites and developments in remote sensing technology, the resolution, type, and quality of remote sensing data improved significantly. In the 21st century, the emergence of high-resolution commercial satellites and multi-spectral imaging technology, the integration of remote sensing technology with computer technology, GIS (geographic information systems), and the Internet have led to methodological innovations in map production, and the extraction of information from remote sensing data has become more efficient and precise, making

DOI: 10.1201/9781003486060-18

remote sensing data even more prominent in map production. Although these methods are highly automated and efficient, the obtained map elements are limited in their planimetric accuracy and do not have accurate elevation information due to image resolution and quality limitations.

In recent years, mobile measurement systems equipped with various sensors (including laser scanners, GNSS (Global Navigation Satellite System), IMU (Inertial Measurement Units)) can simultaneously acquire high-precision point cloud of the surface of roads and roadside features, which contains rich 3D geometric and attribute information (e.g., reflective strength, number of echo counts), and provides an important source of data for the understanding of the road scene, high-precision 3D reconstruction, and HD map production. Based on the property that the highly reflective material in road traffic signs and markings makes their reflective intensity attributes in the point cloud much higher than those of the surrounding roadway points, point cloud data have been widely used in HD map production. By extracting the intensity features of road scene point clouds to extract lane lines and traffic signs, the three-dimensional spatial location of lanes and the topological relationship between lanes can be accurately depicted. Figure 13.1 shows the HD map results extracted from the point cloud. In addition, by mining the 3D spatial information embedded in the point cloud, lane attributes such as lane width, slope, and curvature can be accurately calculated. Figure 13.2 shows the alignment extraction and 3D modeling results using the point cloud in highways.

13.2 FOREST CARBON SINK

Close monitoring of carbon sinks is critical to researching global climate change, and the current status and potential absorption capability of the carbon sinks in terrestrial ecosystems urgently need further studies (Ding 2021). Forests, grasslands, shrubs, crops, and soil are the main carbon pools in terrestrial ecosystems (Fang et al., 2007). The United Nations Food and Agriculture Organization (FAO) pointed out that globally, the carbon stocks of forests account for about 77% of the total carbon stocks in vegetation, making them the most important carbon pool in terrestrial ecosystems (Canton 2021). However, there are significant differences in current estimations of forest carbon sinks based on inventory, modeling, and atmospheric inversion methods.

The issue of how to achieve reliable monitoring of global forest carbon sinks is of great urgency. With the increasing refinement of satellite remote sensing images in space, time, and spectral domains, the growing number of satellite laser data acquisition platforms, and the launching of numerous SAR sensors for forestry and ecological monitoring missions, using spectral information from satellite remote sensing images and vertical forest structure information from airborne/spaceborne LiDAR, in combination with ground observation data to monitor the dynamics of global/regional forest carbon sinks has become a hot topic.

Based on the point cloud data, an interpretable and precision controllable forest carbon stock calculation method is built, which uses the resolution, forest canopy density, terrain slope, and forest height of each pixel to express carbon stock C explicitly. Two Chinese fir forests located in Rongshui and Rong'an Counties, Guangxi, China, are used for testing. As shown in Figure 13.3, the whole region is

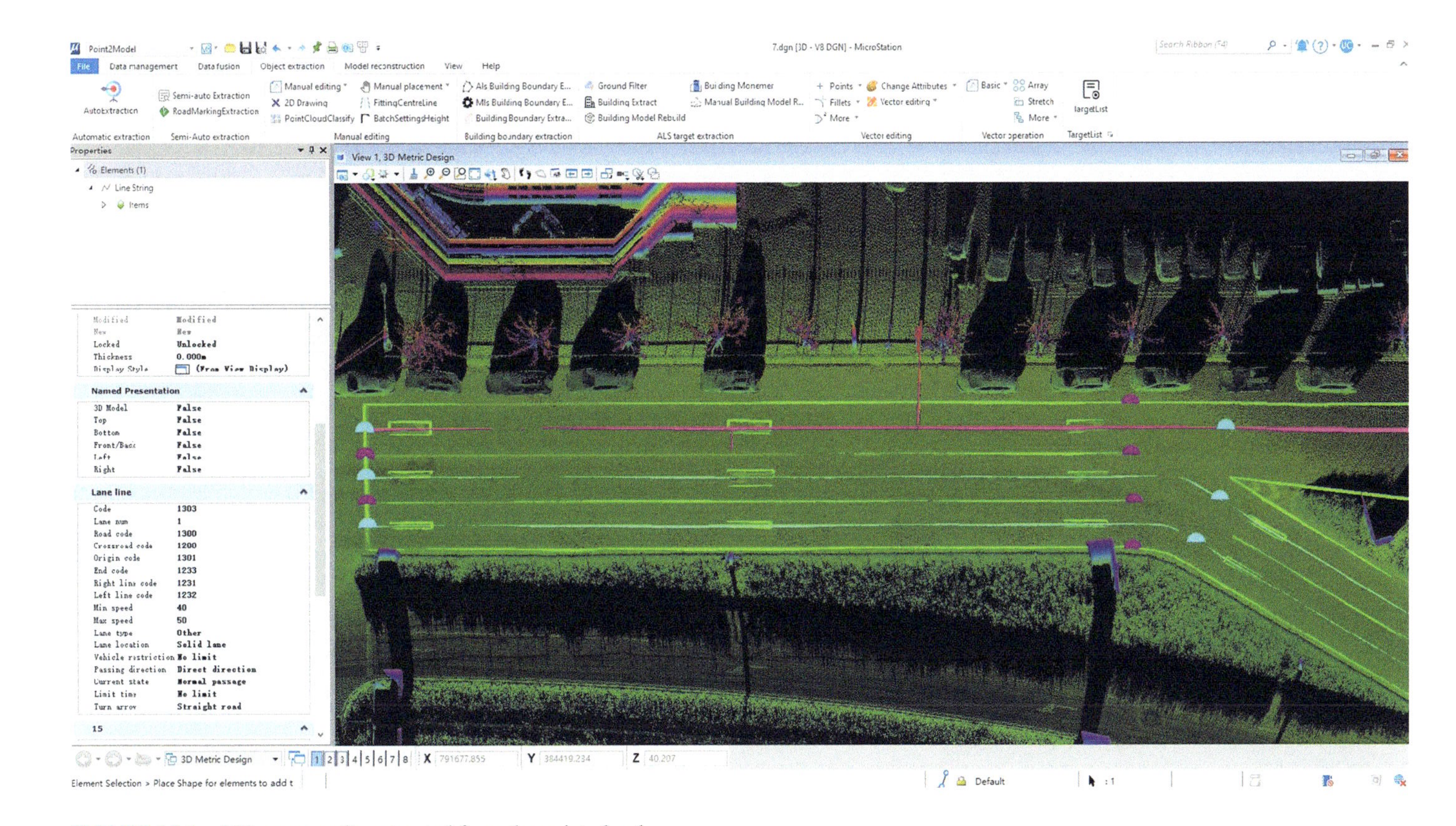

FIGURE 13.1 HD map results extracted from the point cloud.

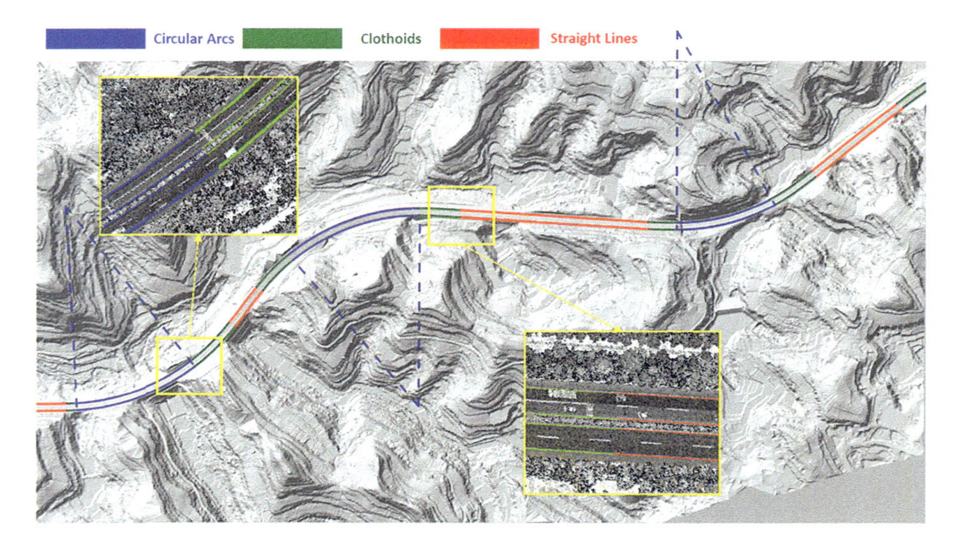

FIGURE 13.2 Alignments extraction and 3D modeling results in highways.

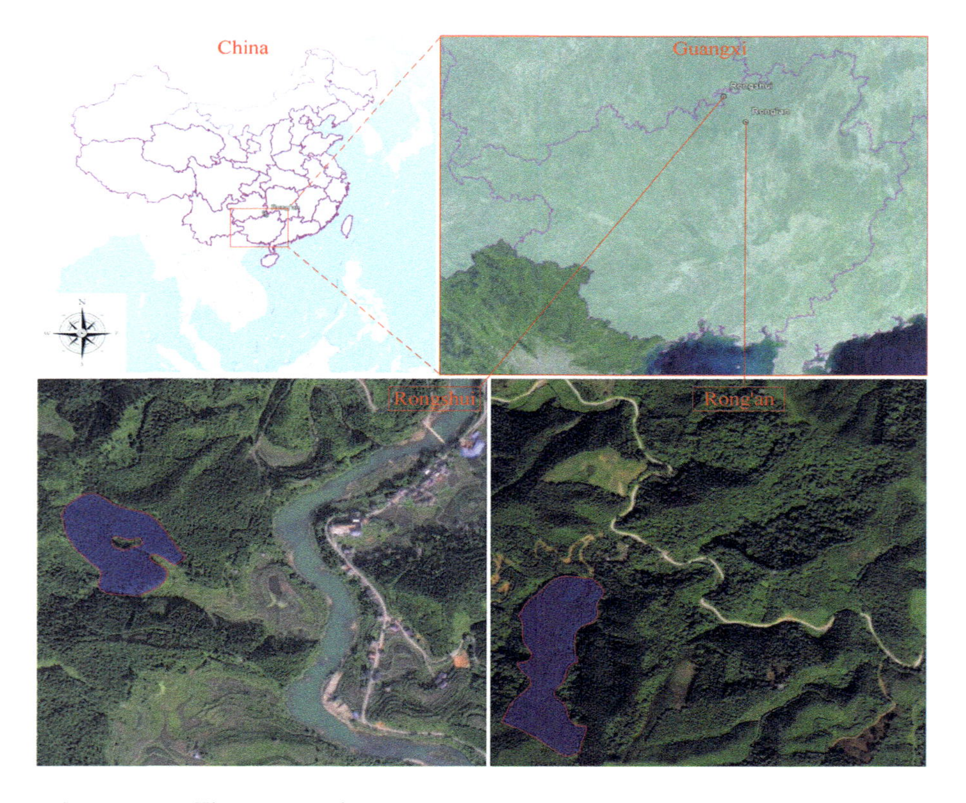

FIGURE 13.3 The test area of Rongshui and Rong'an.

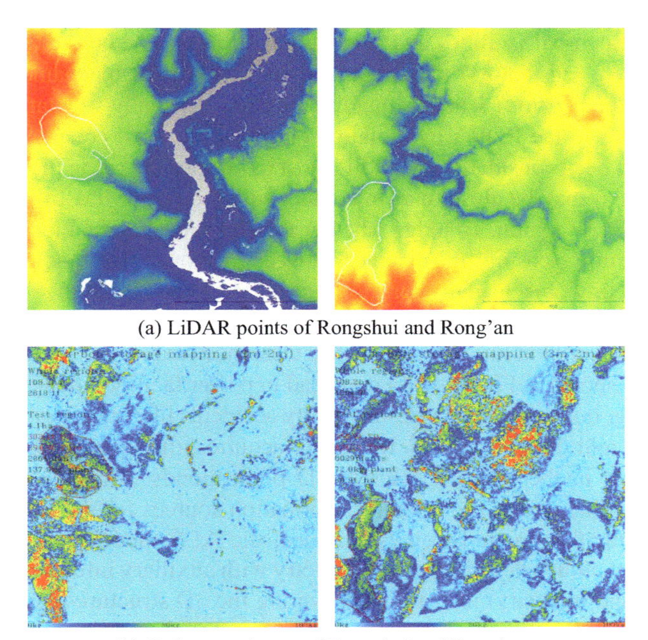

(a) LiDAR points of Rongshui and Rong'an

(b) Carbon stock map of Rongshui and Rong'an

FIGURE 13.4 Carbon stock maps (resolution: 2m × 2m).

about 1km². The two blue validation areas are mountainous terrain, the Rongshui validation area is 4.1ha, including 2864 trees, and the Rong'an validation area is 4.8ha, including 6029 trees. For each validation area, the D and H of all Chinese firs inside were measured manually, the carbon stock of each tree was calculated using a tree-level model, and the total carbon stock was treated as the true value.

By using laser scanning, the map of forest carbon stock with high resolution and accuracy is rapidly generated. Figure 13.3 shows the saturation of forests is very high; Figure 13.4(a) shows the airborne LiDAR points, with airborne LiDAR capable of penetrating forest canopies, high-precision DEM and DSM were obtained by filtering the points, then parameters were calculated with high accuracy. Figure 13.4(b) shows a high precision mapping of forest carbon stock with 2m resolution, and the total carbon stocks are 2618.1 mg and 4294.9 mg respectively.

13.3 ELECTRICITY POWER CORRIDOR INSPECTION

Electricity powers the activities of modern-day societies. The power line networks transmitting the electricity are located in numerous transmission line corridors (i.e., Right-of-Way (ROW)) areas. Inspecting the ROWs is one of the major responsibilities of operating and maintaining power transmission grids. ROWs are vulnerable to potential hazards (e.g., icing), resulting in short- or long-term power outages. The safe operation of power transmission grids must monitor the ROW to find potential threats on the power line networks efficiently and reliably. Inspecting the ROWs incorporates two aspects: electrical facilities and surrounding objects. The condition of the power components (e.g., insulators) must regularly be checked for mechanical and electrical faults (e.g., corrosion). The surrounding objects, especially vegetation, often pose potential threats to the ROWs by violating clearance between conductors and assets. Vegetation has been considered one of the most hazardous factors in the ROW environments because it may be close to or in contact with the power line by growing in and falling in the ROW region. To judge whether the clearance is safe, the ROW clearance anomaly detection measures the distance between the power lines and the surrounding non-power-facility objects in the corridor, such as trees and buildings. The ROW clearance management plays a crucial role in power line risk management and is an important task in transmission line inspection. The clearance hazard (e.g., tree encroachment discharges, as shown in Figure 13.5) may lead to power facility infrastructure failures, power-tripping, outages, or even serious safety accidents like bushfires.

LiDAR (Light Detection and Ranging) acquires the spatial geometry of the detected object in the form of discrete 3D point clouds. Airborne laser scanning (ALS) has become one of the primary information sources for the ROW inspection. Compared with the image-based techniques, the ALS method directly generates dense 3D point clouds describing the geometry with auxiliary information (e.g., multiple echoes and intensity) instead of recovering the 3D structure from stereo vision (Matikainen et al., 2016). Various approaches have been reported on the ROW point clouds classification and the subsequent power facilities modeling including power lines and pylons. General machine learning methods, for example, Dempster–Shafer theory, supervised classification using Random Forest classifier and JointBoost classifier, have been used to accomplish the probabilistic prediction of ROW class label (e.g., power line) (Clode and Rottensteiner, 2005; Kim and Sohn, 2013; Guo et al., 2015). Substantial training samples are required in machine learning methods, and the selection of training samples can influence the classification accuracy.

ALS data is suitable for the ROW inspection. However, the standard manned aircraft-borne (e.g., helicopter) ALS LiDAR systems are limited because of the inflexibility and high costs of flight acquisition. Moreover, the spatial and temporal resolutions of fixed-wing aircraft-borne ALS are relatively low due to the long

(a) Photo taken at the clearance anomaly location.

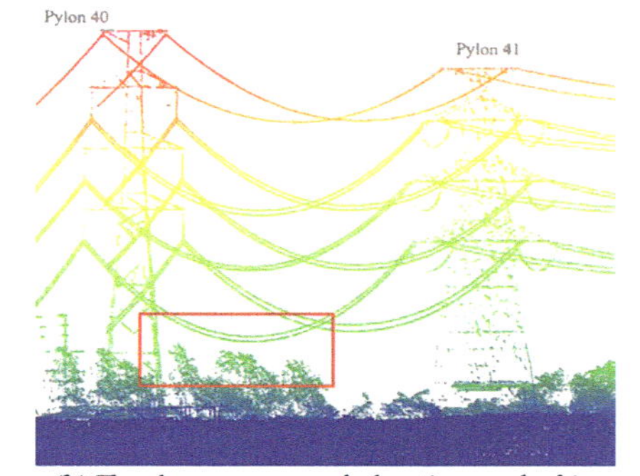

(b) The clearance anomaly location marked in LiDAR point clouds.

FIGURE 13.5 Electricity power corridors invaded by vegetation (Chen et al., 2018).

observation distance. Unmanned aerial vehicles (UAVs) provide an alternative to traditional remote sensing platforms. In recent years, UAV LiDAR systems equipped with GPS, IMU, laser scanners, and optical cameras have presented several characteristics, including flexibility, high resolution, efficiency, and potential for customization, and have led to promising solutions for many applications including fine-scale mapping, biomass change detection, and forest inventory (Wallace et al., 2012; Lin, Hyyppä, and Jaakkola 2010; Jaakkola et al., 2010). UAVs are flexible and can acquire RS data at low altitudes (e.g., 100 m). UAV LiDAR complements standard remote ALS acquisition systems in both spatial and temporal resolutions because of the flexible and close-range data collection. Recent examples of UAV LiDAR systems are operated by a field operator (B. Yang and Chen 2015; Jaakkola et al., 2010). However, with the perfection of the related legal regulations and technological developments of the full autonomous navigation of the UAV, manual interventions from pilots can be downgraded, resulting in further efficiencies. Using UAV LiDAR data to detect the ROW clearance anomaly can effectively improve the flexibility and mobility of the power line inspection and save a large number of manpower, material resources, and time costs in the manned aerial vehicle or manual inspection. It is particularly important for transmission line inspection in difficult terrains, such as mountainous regions (Chen et al., 2018). An example of classified point clouds from UAV LiDAR data is shown in Figure 13.6.

Deep learning greatly improves the performance of almost all recognition tasks in computer vision, such as segmentation, edge detection, and object detection. However, there are very few studies on designing deep learning methods for the real-time understanding of the power corridor scenes (J. Yang et al., 2019). Existing studies are implemented mostly using rule-based or traditional machine learning techniques (Chen et al., 2018; Sohn, Jwa, and Kim 2012; Gu, Wang, and Xie 2017; Kim and Sohn 2013). There are certain limitations that need to be solved:

1. The lack of timeliness. Most previous studies implement the extraction in an offline post-processing manner or with a serial algorithm design that is hard to compute on GPUs parallelly (Pu et al., 2019). Also, the direct processing/

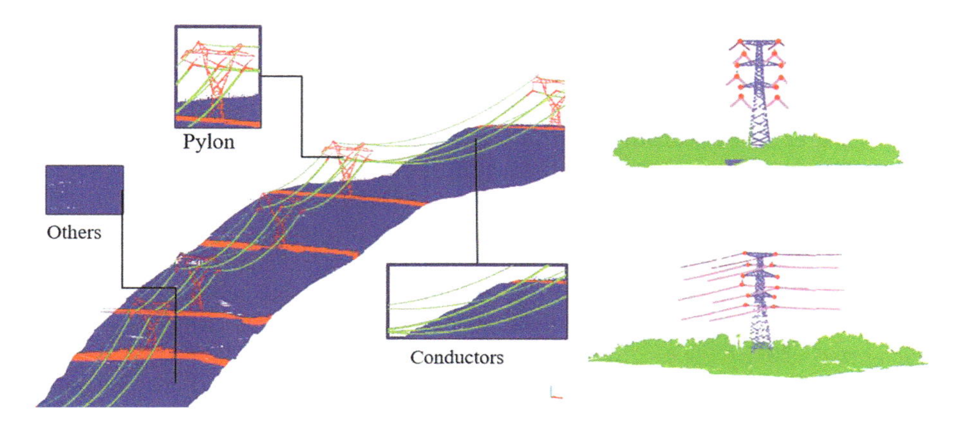

FIGURE 13.6 Classified electricity power corridor point clouds.

learning on raw point clouds instead of reconfigurable computing gird/voxels would complex the algorithm and limit its adaptation on variable GPU accelerated parallel computing devices, especially on the embedded edge computing devices such as the NVIDIA Jetson;
2. The generalizability is problematic. There are various kinds of conductors. Rule-based methods can hardly detect conductors not in the pre-defined rulesets. The point count on the conductors is extremely small compared with the other categories (e.g., ground) in the ROW. The distribution of the training/validation/ test samples is long-tailed. Most existing learning-based methods either overlooked this phenomenon or were trained and validated with small samples.

One of the current state-of-the-art approaches is DCPLD-Net: a diffusion-coupled convolution neural network for real-time power transmission lines detection from UAV-borne LiDAR data. To implement efficient 2D convolution on 3D point clouds, DCPLD-Net proposed a novel point cloud representation named Cross-Section View (CSV), which transforms the discrete point clouds into 3D tensors constructed by voxels with a deformed geometric shape along the flight trajectory. After the CSV feature generation, the encoded features of each voxel are treated as energy (i.e., heat) signals and diffused in space to generate diffusion feature maps. The features of the power line points are thus enhanced through this simulated physical process (diffusion). Finally, a single-stage detector named PLDNet was proposed for the multi-scale detection of conductors on the diffused CSV representations. The experimental results show that the DCPLD-Net achieves a significant improvement on ROWs inspection LiDAR datasets collected by both mini-UAV LiDAR and large-scale fully autonomous UAV power lines inspection robots, and surpasses compared methods (i.e., PointNet ++, RandLANet) in terms of F1 scores and IoU (Qi et al., 2017; Hu et al., 2020).

13.4 LAVA TUNNEL MAPPING

For over half a century, numerous countries have successively organized significant scientific and engineering projects in the field of deep space exploration to search for extraterrestrial life and habitable environments for humans (Luan 2016; Wang et al., 2019). Research indicates that there are abundant lava tunnels on the Moon and Mars, characterized by a relatively enclosed environment and stable structure, providing ideal locations for the construction of extraterrestrial exploration bases and serving as important subjects for deep space exploration (Ye et al., 2016; Yu et al., 2016; Li et al., 2018). Therefore, conducting in-depth comparative planetary studies on Earth's lava tunnels is necessary. Through morphological analysis of surface lava tunnels, a deeper understanding of similar terrain features and geological structures on celestial bodies such as the Moon and Mars can be achieved (Xiao et al., 2018; Geng et al., 2018).

However, the interior environment of lava tunnels is enclosed, the terrain is rugged, and the lighting conditions are inadequate, making detecting lava tunnels challenging. Figure 13.7 depicts Xianren lava tunnel and Qishier lava tunnel near the

FIGURE 13.7 Example of lava tunnel: (a) Xianren lava tunnel; (b) Qishier lava tunnel.

Shishan Volcanic Group Geological Park in Haikou, Hainan Province, China. Laser scanning systems have low environmental requirements, are independent of lighting conditions, and provide three-dimensional coordinates of objects in the form of point cloud data, making them an excellent means for detecting lava tunnels. Typically, a mobile measurement platform can be employed, combined with simultaneous mapping and localization algorithms, to obtain a point cloud map of the lava tunnel. Alternatively, multiple ground station scanning systems can be deployed to collect data, followed by offline point cloud registration to obtain a complete dataset of the lava tunnel. The mobile measurement platform offers higher efficiency, while the ground station scanning system provides higher point cloud accuracy.

After completing the production of the point cloud model, a three-dimensional model of the lava tunnel can be obtained through intelligent processing methods for point cloud data, followed by morphological analysis. The formation of lava tunnels is complex and influenced by factors such as topography and magma viscosity. The general evolutionary process involves the overflow of magma during a volcanic eruption, where the surface layer solidifies into a crust first due to radiative cooling. At this point, the internal lava, with slower cooling, retains higher fluidity. When the volcanic eruption weakens, and subsequent lava flow stops, the internal lava flow empties, forming the lava tunnel. Lava tunnels often possess features such as collapse pits, debris piles, skylights, lava stalactites, and lava stalagmites. To accurately model these specific features, the entire lava tunnel can be divided into the ground, tunnel walls, and debris piles. Semantic extraction can then be used to identify specific features. Figure 13.8 illustrates the algorithmic workflow for the three-dimensional modeling and morphological analysis of the lava tunnel.

Firstly, data preprocessing is carried out, removing noise to eliminate the influence of noise points on generating the three-dimensional model in the point cloud map. Subsequently, ground filtering is performed, and connectivity analysis is used to further separate debris piles and tunnel walls. Secondly, after semantic extraction of specific features, normal vector estimation is conducted to complete surface reconstruction. A hole-filling method is employed to supplement collected data and fill spatial gaps that may have occurred during data pre-processing to ensure the integrity of features. Upon obtaining the three-dimensional model of the lava tunnel, the skeleton lines of

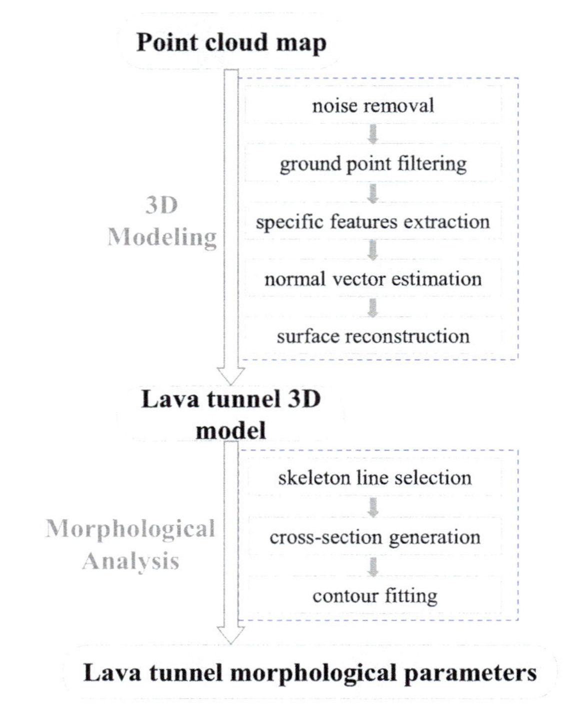

FIGURE 13.8 The algorithm of 3D modeling and morphological analysis of lava tunnel.

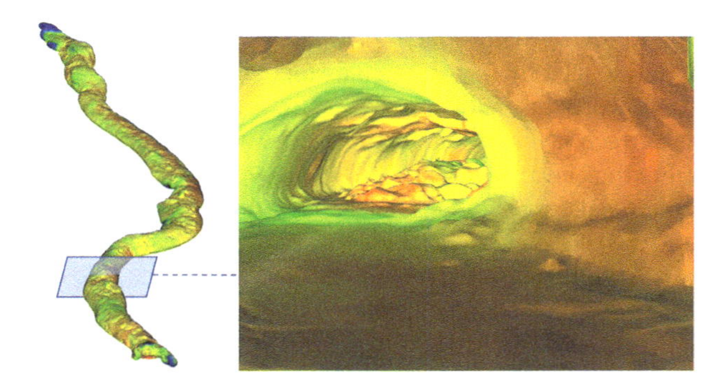

FIGURE 13.9 The 3D model of Xianren lava tunnel.

typical regions (such as corners and significant elevation changes in the lava tunnel) are extracted. Cross-sections are generated using these skeleton lines and the lava tunnel, followed by curve fitting to obtain the cross-sectional profiles. Tunnel parameters are then calculated based on these profiles for morphological analysis.

Figure 13.9 shows the processing results for Xianren lava tunnel. The three-dimensional model and morphological analysis of the lava tunnels provide reliable data support for relevant scholars, which is of great significance.

REFERENCES

Canton, Helen. 2021. "Food and Agriculture Organization of the United Nations—FAO." In *The Europa Directory of International Organizations 2021*, 297–305. Routledge. https://www.taylorfrancis.com/chapters/edit/10.4324/9781003179900-41/food-agriculture-organization-united-nations%E2%80%94fao-helen-canton

Chen, Chi, Bisheng Yang, Shuang Song, Xiangyang Peng, and Ronggang Huang. 2018. "Automatic Clearance Anomaly Detection for Transmission Line Corridors Utilizing UAV-Borne LIDAR Data." *Remote Sensing* 10 (4): 613.

Clode, Simon, and Franz Rottensteiner. 2005. "Classification of Trees and Powerlines from Medium Resolution Airborne Laserscanner Data in Urban Environments." In *Proceedings of the APRS Workshop on Digital Image Computing (WDIC), Brisbane, Australia*, 21: 1.

Ding, Zhongli. 2021. "Research on China's Carbon Neutralization Framework Roadmap." *China Industry and Information Technology* 8: 54–61.

Fang, J. Y., Z. D. Guo, S. L. Piao, and A. P. Chen. 2007. "Estimation of Terrestrial Vegetation Carbon Sinks in China from 1981 to 2000." *Science in China (Series D)* 37 (6): 804–12.

Geng, Y., Z. Jishi, L. I. Sha, F. U. Zhongliang, M. Linzhi, L. I. U. Jianjun, and W. Haipeng. 2018. "A Brief Introduction of the First Mars Exploration Mission in China." *Journal of Deep Space Exploration* 5 (5): 399–405.

Gu, Yanfeng, Qingwang Wang, and Bingqian Xie. 2017. "Multiple Kernel Sparse Representation for Airborne LiDAR Data Classification." *IEEE Transactions on Geoscience and Remote Sensing* 55 (2): 1085–105. doi:10.1109/TGRS.2016.2619384

Guo, Bo, Xianfeng Huang, Fan Zhang, and Gunho Sohn. 2015. "Classification of Airborne Laser Scanning Data Using JointBoost." *ISPRS Journal of Photogrammetry and Remote Sensing* 100: 71–83. doi:10.1016/j.isprsjprs.2014.04.015

Hu, Qingyong, Bo Yang, Linhai Xie, Stefano Rosa, Yulan Guo, Zhihua Wang, Niki Trigoni, and Andrew Markham. 2020. "Randla-Net: Efficient Semantic Segmentation of Large-Scale Point Clouds." In *Proceedings of the IEEE Computer Society Conference on Computer Vision and Pattern Recognition*, 11105–14. doi:10.1109/CVPR42600.2020.01112

Jaakkola, Anttoni, Juha Hyyppä, Antero Kukko, Xiaowei Yu, Harri Kaartinen, Matti Lehtomäki, and Yi Lin. 2010. "A Low-Cost Multi-Sensoral Mobile Mapping System and Its Feasibility for Tree Measurements." *ISPRS Journal of Photogrammetry and Remote Sensing* 65 (6): 514–22. doi:10.1016/j.isprsjprs.2010.08.002

Kim, Heungsik B., and Gunho Sohn. 2013. "Point-Based Classification of Power Line Corridor Scene Using Random Forests." *Photogrammetric Engineering & Remote Sensing* 79 (9): 821–33.

Li, L. J., G. Yan, C. A. O. Jinbin, Z. Tielong, F. Guangyou, Y. Jianfeng, S. H. U. Rong, Z. O. U. Yongliao, and L. I. N. Yangting. 2018. "Scientific Objectives and Payload Configuration of China's First Mars Exploration Mission." *Journal of Deep Space Exploration* 5 (5): 406–13.

Lin, Yi, Juha Hyyppä, and Anttoni Jaakkola. 2010. "Mini-UAV-Borne LIDAR for Fine-Scale Mapping." *IEEE Geoscience and Remote Sensing Letters* 8 (3): 426–30.

Luan, Enjie. 2016. "The Rising China Space Heading for the Endless Universe." *J. Deep Space Explor* 3 (4): 295–306.

Matikainen, Leena, Matti Lehtomäki, Eero Ahokas, Juha Hyyppä, Mika Karjalainen, Anttoni Jaakkola, Antero Kukko, and Tero Heinonen. 2016. "Remote Sensing Methods for Power Line Corridor Surveys." *ISPRS Journal of Photogrammetry and Remote Sensing* 119 (September): 10–31. doi:10.1016/j.isprsjprs.2016.04.011

Pu, S., L. Xie, M. Ji, Y. Zhao, W. Liu, L. Wang, Y. Zhao, F. Yang, and D. Qiu. 2019. "Real-Time Powerline Corridor Inspection by Edge Computing of Uav Lidar Data." In *International Archives of the Photogrammetry, Remote Sensing and Spatial Information Sciences - ISPRS Archives*, 42: 547–51. Copernicus GmbH. doi:10.5194/isprs-archives-XLII-2-W13-547-2019

Qi, Charles R., Li Yi, Hao Su, and Leonidas J. Guibas. 2017. "PointNet++: Deep Hierarchical Feature Learning on Point Sets in a Metric Space." *Advances in Neural Information Processing Systems* 2017: 5100–9.

Sohn, G., Y. Jwa, and H. B. Kim. 2012. "Automatic Powerline Scene Classification and Reconstruction Using Airborne Lidar Data." In *ISPRS Annals of the Photogrammetry, Remote Sensing and Spatial Information Sciences*, 1: 167–72. doi:10.5194/isprsannals-I-3-167-2012

Wallace, Luke, Arko Lucieer, Christopher Watson, and Darren Turner. 2012. "Development of a UAV-LiDAR System with Application to Forest Inventory." *Remote Sensing* 4 (6): 1519–43. doi:10.3390/rs4061519

Wang, Chi, Xianguo Zhang, Xinfeng Xu, and Yueqiang Sun. 2019. "The Lunar and Deep Space Environment Exploration in China." *Journal of Deep Space Exploration* 6 (2): 105–18.

Xiao, Long, Jun Huang, Jiawei Zhao, and Jiannan Zhao. 2018. "Significance and Preliminary Proposal for Exploring the Lunar Lava Tubes." *Scientia Sinica Physica, Mechanica & Astronomica* 48 (11): 119602.

Yang, Bisheng, and Chi Chen. 2015. "Automatic Registration of UAV-Borne Sequent Images and LiDAR Data." *ISPRS Journal of Photogrammetry and Remote Sensing* 101 (March): 262–74. doi:10.1016/j.isprsjprs.2014.12.025

Yang, Jisheng, Zijun Huang, Maochun Huang, Xianxian Zeng, Dong Li, and Yun Zhang. 2019. "Power Line Corridor LiDAR Point Cloud Segmentation Using Convolutional Neural Network." In *Lecture Notes in Computer Science (Including Subseries Lecture Notes in Artificial Intelligence and Lecture Notes in Bioinformatics)*, 11857 LNCS: 160–71. doi:10.1007/978-3-030-31654-9_14

Ye, P., Y. U. Dengyun, S. U. N. Zezhou, and S. Zhenrong. 2016. "Achievements and Prospect of Chinese Lunar Probes." *Journal of Deep Space Exploration* 3 (4): 323–33.

Yu, D., S. U. N. Zezhou, M. Linzhi, and S. H. I. Dong. 2016. "The Development Process and Prospects for Mars Exploration." *Journal of Deep Space Exploration* 3 (2): 108–13.

14 Conclusion and Outlooks

Point clouds directly represent the three-dimensional digitalization of the physical world and have been widely used in scientific research and engineering applications. This book systematically discusses methods for acquiring ubiquitous point clouds, their processing and management, including point cloud fusion and enhancement, 3D information detection and segmentation, LoD (Level of Detail) modeling, and potential applications. It presents initial theoretical methods for intelligent ubiquitous point cloud processing, providing theoretical support and scientific tools for the intelligent handling of ubiquitous point clouds.

With the rapid development of sensors, chips, the Internet of Things (IoT), and platforms, the efficiency and quality of acquiring large-scale point cloud data are continuously improving. The era of city-scale and even global-scale detailed 3D scenes is approaching. Simultaneously, the rapid advancements in edge computing, 5G communication, deep learning, and 3D vision technologies will further enhance the computational power needed for big data processing.

The theoretical methods for intelligent point cloud processing need further expansion to serve more applications. In the future, the following areas warrant deeper exploration and research.

14.1 UNMANNED AUTONOMOUS SYSTEMS

14.1.1 Collaborative Systems

In recent years, the field of air-ground collaboration and cross-domain exploration has witnessed significant advancements, fueled by rapid technological progress and the growing need for integrated solutions in various domains. Current research has demonstrated the potential of collaborative systems involving both aerial and ground-based assets, showcasing enhanced efficiency, flexibility, and adaptability in diverse applications. A key focus has been on developing seamless communication and coordination between unmanned aerial vehicles (UAVs) and ground-based robotic systems. Incorporating sensor networks, real-time data sharing, and advanced algorithms has improved situational awareness and decision-making capabilities. The integration of artificial intelligence and machine learning techniques has further enhanced the autonomy of these systems, empowering them to adapt to dynamic and complex environments with greater dexterity. Moreover, research has delved into optimizing resource allocation and task distribution, ensuring effective utilization of aerial and ground assets. This has been particularly evident in fields such as disaster response, precise agriculture, and surveillance, where collaborative efforts have demonstrated increased speed and accuracy in data collection and analysis.

DOI: 10.1201/9781003486060-19

14.1.2 Challenges and Promising Prospects

Despite the remarkable advancements, several challenges persist in the domain of air-ground collaboration. A prominent issue is the interoperability between different platforms and communication protocols. Standardization across the diverse range of aerial and ground systems remains a hurdle, limiting seamless integration and hindering the full potential of collaborative efforts. Another significant challenge lies in integrating air-ground collaboration within the observation scale, which brings forth a unique set of challenges that significantly impact the fusion of cross-view data. Differing altitudes, sensor configurations, and field-of-view characteristics create complexities in aligning and integrating data.

Looking ahead, the future of air-ground collaboration holds promising prospects. Addressing current challenges requires concerted efforts toward standardization and developing secure communication protocols. Research should concentrate on devising adaptable frameworks that can accommodate diverse platforms and enable effective interoperability. Integrating advanced technologies such as edge computing is poised to revolutionize air-ground collaboration. Enhanced connectivity and reduced latency will enable real-time decision-making, making these systems more responsive and versatile in dynamic environments. Furthermore, research should delve into innovative approaches to enhance collaborative systems' autonomy and learning capabilities. Advancements in swarm intelligence and human-machine collaboration will foster systems adept at navigating intricate scenarios and adapting to unforeseen challenges.

14.1.3 Applications and Future Potentials

Air-ground collaboration finds application across a spectrum of scenarios, each benefitting from the unique capabilities offered by combined aerial and ground-based systems. In disaster response, for instance, the expeditious deployment of UAVs for aerial surveys combined with ground-based robotic systems for search and rescue operations can significantly improve overall response times and effectiveness. In precision agriculture, the integration of aerial sensing and ground-based robotic actuators facilitates targeted and efficient resource management. Collaborative systems are capable of monitoring crop health from above and executing localized interventions on the ground, thereby optimizing agricultural practices. In security and surveillance, air-ground collaboration enhances monitoring capabilities by integrating the aerial vantage point with ground-level insights. Additionally, infrastructure inspection and maintenance benefit from the synergy between aerial and ground-based assets. UAVs can undertake preliminary assessments and pinpoint potential issues, while ground-based robots are deployed for precise and detailed inspections and repairs.

In conclusion, the future of air-ground collaboration holds great promise, provided that ongoing challenges are addressed. As technology continues to evolve, the seamless integration of aerial and ground-based systems will undoubtedly revolutionize various domains, offering more efficient, adaptable, and collaborative solutions.

14.2 LLM WITH POINT CLOUDS

14.2.1 Large Language Model (LLM)

Language Models refer to deep models that are designed and trained to understand and generate human-like text, playing a core role in natural language processing (NLP). Recently, benefitted from *pre-training technologies* (Devlin, 2018), *transformer architectures* (Vaswani et al., 2017), and *missive training data* (Raffel et al., 2020), Large Language Model (LLM) can learn the patterns and nuances of language and can be fine-tuned for various language-related tasks, such as text generation, translation, completion, summarization, question-answering, and more (Kasneci et al., 2023). One notable example of a Large Language Model is the close-source ChatGPT developed by OpenAI, which has achieved excellent application results with billions of parameters. Additionally, there are several open-source projects that have achieved performance close to ChatGPT and are extensively utilized by researchers, including LLaMA (Touvron et al., 2023), Vicuna (Chiang et al., 2023), etc. A leaderboard of these LLMs can be found in https://huggingface.co/spaces/lmsys/chatbot-arena-leaderboard.

14.2.2 Large Language Model with Images

The tremendous success of LLMs in the field of text processing has greatly encouraged researchers to explore how LLMs handle other modality data (Wu et al., 2023, Panagopoulou et al., 2023, Zhang et al., 2024a, 2024b) as input and output. A predominant area of focus has been the integration of the image modality with LLMs.

Large Vision Language Models (LVLM) (Li et al., 2023, Zhu et al., 2023a, 2023b, Liu et al., 2024a, Liu et al., 2024b, Xiao et al., 2024) empower LLMs to comprehend images, that is, to *use images as input* for LLM to achieve image-based visual tasks, such as visual question answering, captioning and multi-turn conversation. The architecture of LVLM comprises three critical components: the *Visual Encoder*, the *visual-text Adapter*, and the *LLM backbone*. Specifically, images are processed through the *Visual Encoder*, generally CLIP (Radford et al., 2021) or EVA (Fang et al., 2023), to construct visual features. Subsequently, these visual features are projected to the text representation space required by LLM using the *Adapter*, which may be designed as a linear layer or MLP (Liu et al., 2024a), a Q-former (Li et al., 2023), Cross-Attention layers (Xiao et al., 2024), or a tiny LLM (Xiao et al., 2024). LVLM is trained with task-specific vision-language datasets, where the *Encoder* and *LLM* are typically frozen or fine-tuned using parameter-efficient fine-tuning (PEFT) techniques, such as LoRA (Hu et al., 2021).

Multimodal-LLM (MLLM) (Wu et al., 2023, Panagopoulou et al., 2023, Zhang et al., 2024b) tasks a step forward on LVLM, concentrating on visual tasks requiring image outputs other than text. For fine-grained visual tasks such as visual grounding, patch-level or pixel-level visual features are required and fed to LLM with feature position encoding. The LLM embeddings are then directed to a task-specific decoder for visual editing on the input image (Lai et al., 2024). In tasks demanding visual generation, LLM is connected with image generation models (Dhariwal and Nichol, 2021) with an injected projection layer, allowing conversations with both text and images.

14.2.3 LLM WITH POINT CLOUD

Similar to Image-related LLM, point-cloud-related LLM (PLLM) integrates LLM with point cloud serving as input or output (Ma et al., 2024). As shown in Figure 14.1, recent researches focus on designing point cloud encoder for embedding geometric and layout information into point cloud features (Chen et al. 2024, Huang et al., 2023a, Yin et al., 2023), training small adaptors for aligning point cloud features to text feature spaces (Hong et al., 2023, Huang et al., 2023b), collecting and annotating large scale point cloud-language datasets (Zhu et al., 2023b, Deitke et al., 2024, Chen et al., 2020, Jia et al., 2024) specifically for tasks including object captioning (Qi et al., 2024), scene captioning (Chen et al., 2024), dense captioning (Fu et al., 2024), 3D grounding (Hong et al., 2023), 3D (Situated-) QA (Huang et al., 2023b), dialogue (Li et al., 2024), planning (Hong et al., 2023), navigation (Hong et al., 2024), manipulation (Zhang et al., 2024a), etc.

However, there still exist several problems:

- Firstly, *Encoder* of object-level point clouds is well developed as the emergence of millions of annotated object-level point clouds (Deitke et al., 2024, Luo et al., 2024). However, encoder processing scene-level point clouds is still far from meeting requirements due to the lack of scene-level large-scale dataset (Jia et al., 2024) and well-studied pre-training strategies (Chen et al., 2024, Jia et al., 2024).
- Secondly, the scale of task-specific point cloud-language dataset remains insufficient, impeding the well-training of point cloud-language *adapter* (Hong et al., 2023).
- Thirdly, a pivotal challenge lies in how to effectively integrate the unique 3D structure and distribution information of point clouds into *LLM* processing.
- Fourthly, the domain of text (LLM)-driven point cloud editing and generation is still nascent, falling short of being applicable and supportive for point cloud-based MLLM.

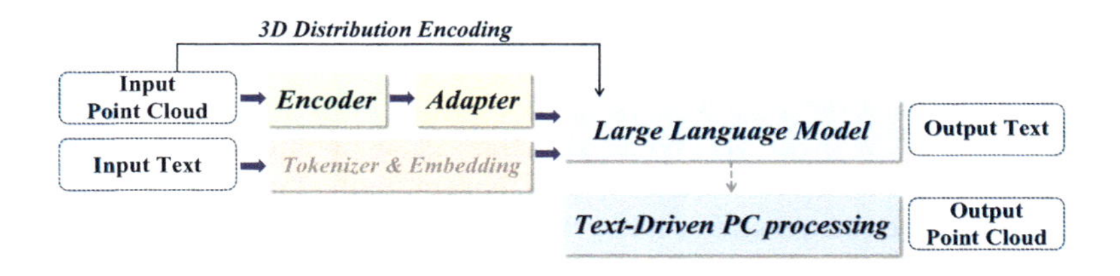

FIGURE 14.1 LLM with point cloud.

14.3 NEW EQUIPMENT FOR POINT CLOUD ACQUISITION

Point cloud acquisition has established a diverse array of multi-platform and multi-resolution data collection systems, encompassing satellite-borne, airborne, vehicle-borne, ground, backpack, and hand-held. Moving forward, the evolution of new equipment is anticipated to be characterized by being lightweight, compact, multi-functional and efficient, which is mainly reflected in the following aspects: (1) Miniaturized multi-spectral laser radar, which will change from geometry-based to simultaneous acquisition of geometry and spectrum/texture in terms of point cloud acquisition content; (2) Single-photon/multi-photon laser radar, which will change from scanning 3D imaging to array single-photon/quantum 3D imaging in terms of imaging mode, improving 3D imaging efficiency; (3) Light and small UAV blue-green laser radar, which will solve the problem of 3D point cloud acquisition in underwater, island and reef areas.

14.4 NEW THEORY OF POINT CLOUD INTELLIGENCE

By drawing on technologies such as artificial intelligence, compressed sensing, machine vision, the Internet of Things, and error processing, we will further expand the data model, processing model, and expression model for intelligent processing of point cloud big data. Technologies such as 5G transmission, edge/cloud computing, and blockchain will enhance the cloud storage and computing capabilities of point cloud big data. Artificial intelligence methods such as deep learning, reinforcement learning, and transfer learning, can bolster the information extraction and expression capabilities of point cloud processing. To solve the core problems of storage and updating, registration and fusion, classification and extraction, expression and analysis confronted by point cloud big data, the new theory will provide core theoretical support for dynamic modeling and real-time simulation analysis of three-dimensional geographic information, thereby building a bridge between point cloud and geoscientific applications.

14.4.1 Point Cloud Storage and Updating

Driven by information technologies, such as cloud storage and cloud computing, we should strengthen the construction of the "cloud-edge-end" collaborative computing framework that integrates cloud storage centers with edge computing of terminal nodes, alleviates the burden on cloud storage centers, and provides robust "computing power" support for two- and three-dimensional integrated computing, analysis, display, and visual decision-making processes. The framework effectively overcomes the current difficulties faced by management and collaborative computing of massive data. Consequently, it is essential to devise a point cloud data structure suitable for cloud storage, rapid visualization, and autonomous updating; investigate methods for detecting changes within point clouds; facilitate the coordinated coupling between point cloud storage, computing, and visualization analysis; realize the classified storage and dynamic update of point clouds, and finally enhance the value of point clouds.

14.4.2 Point Cloud Deep Learning

Point cloud scene interpretation is one of the cores of point cloud intelligence. In-depth exploration is required for point cloud semantic identification, panoptic segmentation, and instance extraction. Current point cloud deep learning methodologies exhibit limitations in handling large-scale scenes and achieving high extraction accuracy. Consequently, it is imperative to devise a point cloud deep learning network structure that meets the above requirements and demonstrates robustness against variations in point cloud density, scene scale, target description completeness, and scene portability. Additionally, the establishment of a representative point cloud scene library is essential to facilitate the extraction of high-level knowledge from point cloud scenes. The deep learning of point clouds should be designed to be more generalizable, particularly in the context of three-dimensional city-level modeling.

14.5 POINT CLOUD BENCHMARK DATASETS

Applying public datasets to evaluate the performance and robustness of research methods is a crucial avenue for advancing research in this domain. However, the existing datasets often fall short in meeting the diverse requirements of various research in terms of data scale, scene diversity, and annotation precision. In terms of public datasets for multi-source point cloud registration, it is essential to further improve the dataset scale, the variety of scene types, the diversity of the platforms, and the multiplicity of acquisition equipment. Recently, the author's research group has released several public datasets. The WHU-TLS dataset for ground point cloud registration contains more than 1.74 billion points in 11 scene types, including urban subways, high-speed railways, mountains, forests, parks, campuses, residential areas, river banks, ancient buildings, underground caves, and tunnels, providing high-quality experimental data for relevant researchers (Dong et al., 2020). In terms of point cloud instances and semantic segmentation, the richly annotated WHU-Urban3D provides extensive coverage of both Airborne Laser Scanning and Mobile Laser Scanning point clouds, along with panoramic images, containing large-scale road and urban scenes in different cities with richly point-wise semantic instance labels (over 200 million points) (Han et al., 2024). Leveraging these datasets, relevant researchers can conduct more effective comparative analyses and methodological evaluations on a consistent basis.

REFERENCES

Chen, Dave Zhenyu, Chang Angel X, and Nießner Matthias. 2020. "Scanrefer: 3D Object Localization in Rgb-d Scans Using Natural Language." In *European Conference on Computer Vision*, 202–21. Springer.

Chen, Sijin, Chen Xin, Zhang Chi, Li Mingsheng, Yu Gang, Fei Hao, Zhu Hongyuan, Fan Jiayuan, and Chen Tao. 2024. "LL3DA: Visual Interactive Instruction Tuning for Omni-3D Understanding Reasoning and Planning." In *Proceedings of the IEEE/CVF Conference on Computer Vision and Pattern Recognition*, 26428–38.

Chiang, Wei-Lin, Zhuohan Li, Zi Lin, Ying Sheng, Zhanghao Wu, Hao Zhang, Lianmin Zheng et al. 2023. "Vicuna: An open-source chatbot impressing gpt-4 with 90%* chatgpt quality." 2(3), 6. See https://vicuna.lmsys.org (accessed 14 April 2023).

Deitke, Matt, Liu Ruoshi, Wallingford Matthew, Ngo Huong, Michel Oscar, Kusupati Aditya, Fan Alam, et al. 2024. "Objaverse-Xl: A Universe of 10m+ 3D Objects." *Advances in Neural Information Processing Systems* 36, 35799–813.

Devlin, Jacob. 2018. "Bert: Pre-training of deep bidirectional transformers for language understanding." arXiv preprint arXiv:1810.04805.

Dhariwal, Prafulla, and Alexander Nichol. 2021. "Diffusion models beat gans on image synthesis." *Advances in Neural Information Processing Systems* 34, 8780–8794.

Dong, Zhen, Fuxun Liang, Bisheng Yang, Yusheng Xu, Yufu Zang, Jianping Li, Yuan Wang et al. "Registration of large-scale terrestrial laser scanner point clouds: A review and benchmark." *ISPRS Journal of Photogrammetry and Remote Sensing* 163 (2020): 327–42.

Fang, Yuxin, Wen Wang, Binhui Xie, Quan Sun, Ledell Wu, Xinggang Wang, Tiejun Huang, Xinlong Wang, and Yue Cao. 2023. "Eva: Exploring the limits of masked visual representation learning at scale." In *Proceedings of the IEEE/CVF Conference on Computer Vision and Pattern Recognition*, pp. 19358–19369.

Fu, Rao, Liu Jingyu, Chen Xilun, Nie Yixin, and Xiong Wenhan. 2024. "Scene-Llm: Extending Language Model for 3D Visual Understanding and Reasoning." *arXiv Preprint arXiv: 2403.11401.*

Han, Xu, Chong Liu, Yuzhou Zhou, Kai Tan, Zhen Dong, and Bisheng Yang. "WHU-Urban3D: An urban scene LiDAR point cloud dataset for semantic instance segmentation." *ISPRS Journal of Photogrammetry and Remote Sensing* 209 (2024): 500–13.

Hong, Yining, Zhen Haoyu, Chen Peihao, Zheng Shuhong, Du Yilun, Chen Zhenfang, and Gan Chuang. 2023. "3D-LLM: Injecting the 3D World into Large Language Models." *Advances in Neural Information Processing Systems* 36: 20482–94.

Hong, Yining, Zheng Zishuo, Chen Peihao, Wang Yian, Li Junyan, and Gan Chuang. 2024. "MultiPLY: A Multisensory Object-Centric Embodied Large Language Model in 3D World." *arXiv Preprint arXiv:2401.08577.*

Hu, Edward J., Yelong Shen, Phillip Wallis, Zeyuan Allen-Zhu, Yuanzhi Li, Shean Wang, Lu Wang, and Weizhu Chen. 2021. "Lora: Low-rank adaptation of large language models." arXiv preprint arXiv:2106.09685.

Huang, Haifeng, Wang Zehan, Huang Rongjie, Liu Luping, Cheng Xize, Zhao Yang, Jin Tao, and Zhao Zhou. 2023a. "Chat-3D v2: Bridging 3D Scene and Large Language Models with Object Identifiers." *arXiv Preprint arXiv:2312.08168.*

Huang, Jiangyong, Yong Silong, Ma Xiaojian, Linghu Xiongkun, Li Puhao, Wang Yan, Li Qing, Zhu Song-Chun, Jia Baoxiong, and Huang Siyuan. 2023b. "An Embodied Generalist Agent in 3D World." *arXiv Preprint arXiv:2311.12871.*

Jia, Baoxiong, Chen Yixin, Yu Huangyue, Wang Yan, Niu Xuesong, Liu Tengyu, Li Qing, and Huang Siyuan. 2024. "SceneVerse: Scaling 3D Vision-Language Learning for Grounded Scene Understanding." *arXiv Preprint arXiv:2401.09340.*

Kasneci, Enkelejda, Seßler Kathrin, Küchemann Stefan, Bannert Maria, Dementieva Daryna, Fischer Frank, Gasser Urs, et al. 2023. "ChatGPT for Good? On Opportunities and Challenges of Large Language Models for Education." *Learning and Individual Differences* 103: 102274.

Lai, Xin, Zhuotao Tian, Yukang Chen, Yanwei Li, Yuhui Yuan, Shu Liu, and Jiaya Jia. 2024. "Lisa: Reasoning segmentation via large language model." In *Proceedings of the IEEE/CVF Conference on Computer Vision and Pattern Recognition*, pp. 9579–9589.

Li, Junnan, Li Dongxu, Savarese Silvio, and Hoi Steven. 2023. "Blip-2: Bootstrapping Language-Image Pre-Training with Frozen Image Encoders and Large Language Models." In *International Conference on Machine Learning*, 19730–42. PMLR.

Li, Zeju, Zhang Chao, Wang Xiaoyan, Ren Ruilong, Xu Yifan, Ma Ruifei, and Liu Xiangde. 2024. "3DMIT: 3D Multi-Modal Instruction Tuning for Scene Understanding." *arXiv Preprint arXiv:2401.03201.*

Liu, Hao, Yan Wilson, Zaharia Matei, and Abbeel Pieter. 2024a. "World Model on Million-Length Video And Language With RingAttention." *arXiv Preprint arXiv:2402.08268.*

Liu, Haotian, Li Chunyuan, Wu Qingyang, and Lee Yong Jae. 2024b. "Visual Instruction Tuning." *Advances in Neural Information Processing Systems* 36, 34892–916.

Luo, Tiange, Rockwell Chris, Lee Honglak, and Johnson Justin. 2024. "Scalable 3D Captioning with Pretrained Models." *Advances in Neural Information Processing Systems* 36, 75307–337.

Ma, Xianzheng, Bhalgat Yash, Smart Brandon, Chen Shuai, Li Xinghui, Ding Jian, Gu Jindong, et al. 2024. "When LLMs Step into the 3D World: A Survey and Meta-Analysis of 3D Tasks via Multi-Modal Large Language Models." *arXiv Preprint arXiv:2405.10255.*

Panagopoulou, Artemis, Xue Le, Yu Ning, Li Junnan, Li Dongxu, Joty Shafiq, Xu Ran, Savarese Silvio, Xiong Caiming, and Niebles Juan Carlos. 2023. "X-InstructBLIP: A Framework for Aligning X-Modal Instruction-Aware Representations to LLMs and Emergent Cross-Modal Reasoning." *arXiv Preprint arXiv:2311.18799.*

Qi, Zhangyang, Fang Ye, Sun Zeyi, Wu Xiaoyang, Wu Tong, Wang Jiaqi, Lin Dahua, and Zhao Hengshuang. 2024. "Gpt4point: A Unified Framework for Point-Language Understanding and Generation." In *Proceedings of the IEEE/CVF Conference on Computer Vision and Pattern Recognition*, 26417–27.

Radford, Alec, Jong Wook Kim, Chris Hallacy, Aditya Ramesh, Gabriel Goh, Sandhini Agarwal, Girish Sastry et al. 2021. "Learning transferable visual models from natural language supervision." In *International Conference on Machine Learning*, pp. 8748–8763. PMLR.

Raffel, Colin, Shazeer Noam, Roberts Adam, Lee Katherine, Narang Sharan, Matena Michael, Zhou Yanqi, Li Wei, and Liu Peter J. 2020. "Exploring the Limits of Transfer Learning with a Unified Text-to-Text Transformer." *Journal of Machine Learning Research* 21 (140): 1–67.

Touvron, Hugo, Lavril Thibaut, Izacard Gautier, Martinet Xavier, Lachaux Marie-Anne, Lacroix Timothée, Rozière Baptiste, et al. 2023. "Llama: Open and Efficient Foundation Language Models." *arXiv Preprint arXiv:2302.13971.*

Vaswani, Ashish, Noam Shazeer, Niki Parmar, Jakob Uszkoreit, Llion Jones, Aidan N. Gomez, Łukasz Kaiser, and Illia Polosukhin. 2017. "Attention Is All You Need." In *Advances in Neural Information Processing Systems*, edited by I. Guyon, U. Von Luxburg, S. Bengio, H. Wallach, R. Fergus, S. Vishwanathan, and R. Garnett. Vol. 30. Curran Associates, Inc.

Wu, Shengqiong, Fei Hao, Qu Leigang, Ji Wei, and Chua Tat-Seng. 2023. "Next-Gpt: Any-to-Any Multimodal Llm." *arXiv Preprint arXiv:2309.05519.*

Xiao, Junfei, Xu Zheng, Yuille Alan, Yan Shen, and Wang Boyu. 2024. "PaLM2-VAdapter: Progressively Aligned Language Model Makes a Strong Vision-Language Adapter." *arXiv Preprint arXiv:2402.10896.*

Yin, Fukun, Chen Xin, Zhang Chi, Jiang Biao, Zhao Zibo, Fan Jiayuan, Yu Gang, Li Taihao, and Chen Tao. 2023. "ShapeGPT: 3D Shape Generation with A Unified Multi-Modal Language Model." *arXiv Preprint arXiv:2311.17618.*

Zhang, Duzhen, Yu Yahan, Li Chenxing, Dong Jiahua, Su Dan, Chu Chenhui, and Yu Dong. 2024a. "MM-LLMs: Recent Advances in Multimodal Large Language Models." *arXiv Preeprint arXiv:2401.13601.*

Zhang, Sha, Huang Di, Deng Jiajun, Tang Shixiang, Ouyang Wanli, He Tong, and Zhang Yanyong. 2024b. "Agent3D-Zero: An Agent for Zero-Shot 3D Understanding." *arXiv Preprint arXiv:2403.11835.*

Zhu, Deyao, Chen Jun, Shen Xiaoqian, Li Xiang, and Elhoseiny Mohamed. 2023a. "Minigpt-4: Enhancing Vision-Language Understanding with Advanced Large Language Models." *arXiv Preprint arXiv:2304.10592*.

Zhu, Ziyu, Ma Xiaojian, Chen Yixin, Deng Zhidong, Huang Siyuan, and Li Qing. 2023b. "3D-Vista: Pre-Trained Transformer for 3D Vision and Text Alignment." In *Proceedings of the IEEE/CVF International Conference on Computer Vision*, 2911–21.

Index

Pages in *italics* refer to figures and pages in **bold** refer to tables.

N

O

P

R

S

T

U

V

W